The Nature of the Stratigraphical Record

Third Edition

The Nature of the Stratigraphical Record

Third Edition

DEREK V. AGER

*Emeritus Professor of Geology, University College of Swansea,
University of Wales, UK*

JOHN WILEY & SONS

Chichester · New York · Brisbane · Toronto · Singapore

First published in Great Britain 1973 by The Macmillan Press Ltd, and in the
USA by Halsted Press, a Division of John Wiley & Sons, Inc.
Second edition 1981.

Third edition published 1993 by John Wiley & Sons Ltd,
                                 Baffins Lane, Chichester,
                                 West Sussex PO19 1UD, England

Other Wiley Editorial Offices

John Wiley & Sons, Inc., 605 Third Avenue,
New York, NY 10158-0012, USA

Jacaranda Wiley Ltd, G.P.O. Box 859, Brisbane,
Queensland 4001, Australia

John Wiley & Sons (Canada) Ltd, 22 Worcester Road,
Rexdale, Ontario M9W 1L1, Canada

John Wiley & Sons (SEA) Pte Ltd, 37 Jalan Pemimpin #05-04,
Block B, Union Industrial Building, Singapore 2057

Library of Congress Cataloging-in-Publication Data

Ager, D. V. (Derek Victor)
    The nature of the stratigraphical record / Derek V. Ager. — 3rd
ed.
        p. cm.
    Includes bibliographical references and index.
    ISBN 0-471-93808-4 (paper)
    1. Geology, Stratigraphic. I. Title.
QE651.A37   1993
551.7—dc20   92–30173
                                                        CIP

British Library Cataloguing in Publication Data

A catalogue record for this book is available from the British Library

ISBN 0-471-93808-4

Typeset in 11/13pt Palatino by Dobbie Typesetting Ltd, Tavistock, Devon.
Printed and bound in Great Britain by Biddles Ltd, Guildford and King's Lynn.

For
Kitty and **Martin**

Contents

Preface to the First Edition

This is not a textbook or a research treatise. It is—I like to think— an 'ideas book'. It is a commentary on the general pattern of earth history which I hope will be stimulating, if perhaps provocatively so, to all those concerned with geology as a whole rather than as a loose agglomeration of separated specialities. I wrote it because I had to, because the ideas it contains have been fermenting in my brain for years and I had to write them down before I become completely intoxicated. In a sense, it is stratigraphy looked at by a non-stratigrapher. I have always used and taught stratigraphy rather than actually made it, but I can offer the excuse that the non-combatant usually has a clearer picture of the battle than the soldiers actually engaged in the fighting. It seems to me that the conclusions contained in this book are inescapable, if one is not too involved in the minutiae of stratigraphical correlation actually to see them.

No doubt I shall be criticised for some of my generalisations, but I am unrepentant. It is sad if some of my details are wrong or over-simplified, but I am obstinate enough to maintain that the general principles, in which I am chiefly interested, are right. I have tried, as far as possible, to use examples which I have seen for myself and which have impressed me. If the Chalk and the Wenlock Limestone also occur in Ruritania, then I am delighted to hear it, but I have not seen them there, so it has not registered in my simple mind. My motto in this connection is that of the great Joachim Barrande—'*C'est ce que j'ai vu*'.

For the same reason, I have not made great use of the literature. For the most part I have only referred to books and papers when some substantiation seems to be needed for a certain point. I have also made particular use of work by myself or by friends, where I have had the opportunity to discuss the matter or to see it for myself. For the sake of readability, I have omitted most direct references from the text. I apologise to the authors concerned, but a selection of the more relevant papers are listed in an annotated bibliography at the end of each chapter. These are not intended to be exhaustive.

The one great hope I have for this book is that it will stimulate thought and argument, even rage. I think our science would be a lot healthier if we took less for granted. It may be said that I do not relate my thoughts sufficiently to all the exciting new ideas of sea-floor spreading and plate tectonics that have revolutionised our thinking in the last few years. But really these ideas, exciting though they are, do not help very much. My stratigraphical enigmas remain the same wherever the plates are sailing round the earth. In some cases, rearrangements of the continents merely make my problems more difficult. At times I almost feel more willing to put the blame on flying saucers than on floating plates.

I would like to thank my former colleague Dr Peigi Wallace very sincerely for her careful and critical reading of the manuscript and for our many useful discussions over the years. Though we agree on the general principles discussed here, she is in no way responsible for my errors and excesses. Nor are my former colleagues, Professor Gilbert Kelling and Professor Mike Brooks, both of whom very kindly read and criticised this text.

I am very grateful to Mary Pugh and John Uzzell Edwards for drawing the figures, to Stan Osborne and Christine Nicholls for printing my photographs, and to Mrs J. Nuttall, Miss M. Davies and Mrs J. Waring for typing the manuscript with great care.

Finally, let me say a word to those who think that science in general and stratigraphy in particular are solemn matters that should be treated at all times with respect: you had better not read this book; you will find it too light-hearted.

Derek Ager
Swansea, 1972

Preface to the Second Edition

Reactions to the first edition of this book ranged from near ecstasy to something approaching scorn. Fortunately the former were very much in the majority and nothing I have seen, heard or read since has persuaded me to change my somewhat extreme views; in fact I have been encouraged in my prejudices. Everywhere I go I see rapid deposition, frequent breaks and incredible persistence. I write this as I return from a visit to Kenya where I saw between Mombasa and Malindi the (often oolitic) Bajocian limestones forming an escarpment just like my beloved Cotswolds, the chief difference being that the stately beeches are replaced by massive baobabs. In neighbouring Tanzania I saw repeated rushes of conglomerate in the Upper Jurassic along contemporaneous fault lines.

I have received shoals of reprints about storm deposits up and down the column all round the world, I have been invited to half-a-dozen conferences on variations of the theme 'the importance of the rare event'.

My one serious critic concentrated on my apparent disregard for generations of careful stratigraphical nit-picking. No one who works on the European Jurassic as I do can be unaware of the niceties and necessities for what may be called 'microstratigraphy'. In fact I would emphásise what it has shown us about the complexities of the seemingly most straightforward parts of the stratigraphical record. But I must say again that we should step back at times and see the full width of the canvas. That is what I have tried to do.

In this second edition I have corrected a few of the errors, I have introduced further and better examples that have come to my attention in the meantime and I have expanded here and there on the basis of my own subsequent experience and further thoughts.

I thank many friends and correspondents from all over the world who have commented on the first edition. I have tried to include the points they have raised, where appropriate, but I have also tried to keep the book short and therefore, I hope, readable.

Derek Ager
Swansea, 1980

The basic law of stratigraphy is the Law of Supposition
—from *Geological Howlers,* edited by W. D. I. Rolfe (1980)

Preface to the Third Edition

St Matthew wrote 'A prophet is not without honour, save in his own country . . .' I sometimes think that it should read 'An honour is not without profit, save in its own country . . .' There is very little financial gain, however, to be had from writing scientific books, unless one writes very popular standard school texts. This is not one of those, though one American reviewer did kindly say that it was a 'must'. Fortunately I can declare myself oblivious of the market-place philosophy that now pervades my home country. I write because I enjoy writing and I particularly enjoyed writing this book.

I was delighted therefore, after the British publishers had declared the second edition 'out of print', that the American publishers, John Wiley & Sons, immediately agreed to produce this third edition. It has certainly been very popular west of the Atlantic and brought me a number of invitations to preach my heretical creed in the United States and Canada. Indeed, in view of the literature that has been published since the appearance of the first edition, my heretical creed seems to have become the orthodoxy of stratigraphy. We now read repeatedly of rare events and storm deposits, episodic evolution and extra-terrestrial impacts. Catastrophism is back again. An American author was generous enough to say of this book that it 'has profoundly influenced modern stratigraphic thinking'. The '-ic' ending shows that it was an American, and I am sure that the Queen would say 'stratigraphical' if it ever came into her conversation, so I will retain the title as it was and ask my many American friends to excuse the many quaint souvenirs of the 'Old Country' that creep into the text.

I am particularly grateful to my wife, Renée, for helping me with the boring task of preparing the new index.

I have tried to spread my stratigraphical net as widely as possible. I have been fortunate enough to visit 57 different countries. (The number shot up suddenly this year with the addition of Slovenia, Croatia, Bosnia-Herzegovina and Georgia—Joe Stalin's place, not the one you march through—as separate states.) All my journeys have only served to confirm my ideas or, if you wish, my prejudices. My first head of department, Professor H. H. Read, said that 'the best geologist . . . is he who has seen the most rocks'. I cannot claim to be among the best of geologists, but I can claim to have seen a lot of rocks. That is my chief excuse for making so many wild generalisations in this book.

I ask my readers to be tolerant and to read 'Soviet Union' as 'former Soviet Union' and to read 'Yugoslavia' as 'former Yugoslavia'. It is just not worth the effort of finding out if Umsk or Tumsk are now in Tajikstan or Kyrgyzstan. I hope the inhabitants of those splendid lands will excuse me. I understand that this book has been translated into Chinese and Russian. I am not supposed to know about those, in case I expect royalties. I was interested to learn that the opening of Chapter 7 on 'Marxist Stratigraphy and the Golden Spike' has remained unchanged, so perhaps I was ahead of the politicians.

Though I remain a fairly pure-minded academic geologist at heart, I was pleased to be able to record in my book *The Geology of Europe* (1980, McGraw-Hill, Maidenhead, UK, p. 127) that I had forecast the oil wealth of the North Sea long before the oil people had started work there. I also suggested, in a *New Scientist* article on 11 July 1983, that the cause of the Falklands War in 1982 was not the natural inclination of the British islanders to remain British, but the oil potential of the wide continental shelf surrounding those islands. This is now proving correct and the oil companies are moving in. I came to those conclusions from straight, if somewhat eccentric, geological thinking, with no real knowledge of oil geology or of the detailed data concerned in either case. So, even for harsh commercial reasons, there is still a case for the academic geologist and for the sort of theorising that is found here.

The lesson of this book, if there is one, is always to expect the unexpected.

Derek Ager
Swansea, 1992

1
The Persistence of Facies

UPPER CRETACEOUS CHALK

In 1957 I had the good fortune to visit the geologically exciting country of Turkey, to look at some of the local Mesozoic rocks and their faunas. I was taken by a Turkish friend to visit a cliff section in Upper Cretaceous sediments near Şile on the Black Sea coast. In the Turkish literature these were described as white limestones with chert nodules and a strange-sounding list of fossils. But what I in fact saw was the familiar white chalk of north-west Europe with black flints and old fossil friends such as *Micraster* and *Echinocorys*. What I was looking at was identical with the 'White Cliffs of Dover' in England and the rolling plateau of Picardy in France, the quarries of southern Sweden and the cliffs of eastern Denmark. This set me thinking on the themes that are expressed in this book.

Though I thought the above observation worthy of a mini-publication, it might be said that it was not all that surprising. We have long known, of course, that the White Chalk facies of late Cretaceous times extended all the way from Antrim in Northern Ireland, via England and northern France, through the Low Countries, northern Germany and southern Scandinavia to Poland, Bulgaria and eventually to Georgia in the south of the Soviet Union. We also knew of the same facies in Egypt and Israel. My record was merely an extension of that vast range to the south side of the Black Sea.

Similarly, at the other end of the belt, Chalk was later discovered in south-west Ireland (where it must have been noticed by the early surveyors, but they had evidently been too scared of their autocratic director to record such an unlikely phenomenon). Later still it was found covering extensive areas of the sea-floor south of Ireland.

Now this spread of a uniform facies is remarkable enough, but it must also be remembered that chalk is a very unusual sediment: an extremely pure coccolith limestone which is almost unique in the stratigraphical column. Nevertheless, there is even worse to come, for on the other side of the Atlantic in Texas, we find the Austin Chalk of the same age and character, and later Cretaceous chalks (still contemporaneous with the European development) are found in Arkansas, Mississippi and Alabama. And most surprising of all, much farther away still in Western Australia, we have the Gingin Chalk of Late Cretaceous age, with the same black flints and the same familiar fossils, resting—as in north-west Europe—on glauconitic sands.

Unfortunately I have not yet visited Australia, but Dr Andy Gale tells me (personal communication, 1989) that the succession of the Upper Cretaceous in Western Australia is really remarkably like that in Britain. It is virtually identical, for example, with that in Antrim, Northern Ireland. Thus glauconitic sands or *gaize* are followed by glauconitic marls or *mulatto* and then by chalk marl and chalk with flints, containing the same invertebrate genera. He suggests that the fantastic abundance of coccoliths forming the chalk (ten times what it is at present) would have modified the climate dramatically. Coccolith blooms release dimethyl sulphide which causes natural acid rain, increases the albedo reflecting solar radiation and reverses the 'greenhouse effect'.

This fits in with my ideas of Nature as being the biggest polluter, since the new-fangled plants first poured that dreadful element oxygen into our planet's pure atmosphere of methane and carbon dioxide. Indeed the recent eruption of Pinatubo in the Philippines has spread dust around the world and has outdone Sadam Hussain's efforts in setting fire to the oil-wells of Kuwait, which it was feared would have disastrous effects on the atmosphere.

Some general explanation is surely needed for such a wide distribution of such a unique facies as the chalk during a short period of geological time. What is more, there has been no other deposit quite like it either before or since, except perhaps some Miocene chalks which themselves are remarkably widespread: in the western approaches to the English Channel, in Malta, Cyprus and the Middle East and all the way to New Zealand.

But now let us climb slowly down the stratigraphical column to see what other widespread facies we can find.

'URGONIAN' LIMESTONES

Towards the end of early Cretaceous times, at the Barremian/Aptian level, a massive limestone was developed that is usually called the 'Urgonian facies'. This can be seen in Portugal where it forms the karstic, touristic coast west of Estoril, with strange

Figure 1.1 *Urgonian escarpment at Cuenca, south-east Spain (DVA)*

erosional shapes such as the Boca do Inferno (Mouth of Hell). In Spain it weathers into the mushrooms, ships and other weird pinnacles of the Ciudad Encantada (Enchanted City) near where it shades the back streets of Cuenca (Figure 1.1). It then forms the magnificent cliffs at Cassis, east of Marseilles. It dominates the scenery in the outer ranges of the Alps (for example, the towering escarpment of Glandasse above Hannibal's route through the sub-Alps). It caps many of the textbook ridges in the eastern part of the Jura Mountains (Figure 1.2). It is well developed all round the Carpathians, in Czechoslovakia, Poland and Romania and then through the Balkan Mountains of Bulgaria (Figure 1.3) and across the Black Sea to the Crimea and the Caucasus (Figure 1.4).

Typically, the Urgonian limestones are thought of as rudist reef deposits, and locally they show tremendous in-situ growths of these aberrant bivalves (for example, near the tragic city of Guernica in northern Spain). Elsewhere—as in the Jura—corals are more important than rudists and usually it is only a reef limestone in the broadest Gallic sense of that term; that is to say, a massive limestone without bedding planes and commonly recrystallised. It also tends to develop in huge lenses that pass

Figure 1.2   *Urgonian limestone capping anticlinal ridge of Montagne de la Balme near Bellegarde in the French Jura (DVA)*

Figure 1.3 *Urgonian limestone above Dryanova monastery, near Gabrovo, Bulgaria*
(DVA)

laterally into other facies (well seen, for example, in Bulgaria). Since the publication of the first edition of this book, I have seen a remarkably similar rock of the same age—the Cupido Limestone—dominating the Mesozoic scenery in north-east Mexico. A comparable facies also extends down into northern Africa.

The term 'Urgonian' has become almost a dirty word in Cretaceous stratigraphy, for it is not one of the internationally-accepted stage names and it is said to be a diachronous, southern facies. Admittedly it is diachronous, but hardly by more than

Figure 1.4   Urgonian limestone forming escarpment, valley of River Heory, Georgia, USSR (DVA)

half a stage or so, and it is still a valid generalisation to say that a massive limestone is developed over a very wide area towards the end of Early Cretaceous times.

TITHONIAN LIMESTONES

Below the towering Urgonian cliffs of southern Europe, there is commonly a second great escarpment formed by a massive limestone at the top of the Jurassic succession. Again it runs from northern Africa through Spain and up into the Alps. It forms, for example, the cliffs of the Porte de France at Grenoble. It also caps many of the Jura box-folds, where the Cretaceous succession has been eroded away. Here it illustrates the point that it is the deposition of carbonate that is important and not the precise nature of the sediment. Thus in the outer Alps, the Tithonian is commonly in the form of a *Calpionella* limestone, with abundant tintinnids and such otherwise-unusual forms as ammonite aptychi, nautiloid jaws, specialised belemnites and the aberrant brachiopod with the hole in the middle: *Pygope*. In the Jura the Tithonian can be seen in places to be a coral reef limestone, but usually the corals have been obliterated by dolomitisation and dedolomitisation as is commonly the fate of reefs.

It might also be said that the Portland Limestone of southern England is the Tithonian limestone in yet another form, with what is mainly a molluscan fauna of limited diversity. But the term 'Tithonian', though not quite so lacking in respectability as the 'Urgonian', is usually reserved for the carbonate facies of alpine Europe, and for years competed with the 'Volgian' the honour of being the accepted international term for the topmost stage of the Jurassic.

In its reef or reef-like facies, the Tithonian continues round the Alps, the Carpathians, the Balkan Mountains to the Caucasus. Where best developed, for example at Stramberk in Czechoslovakia, it contains a rich and varied fauna including massive compound corals (though the fossils are more obvious

Figure 1.5 Tithonian limestone capping Jurassic succession at the 'Iron Gates', where the Danube flows through the Carpathians between Romania (on the right) and Yugoslavia (on the left) (DVA)

in the collections from many years of quarrying than in the quarries themselves). Whether or not these were in-situ reefs is still a matter of dispute, but the dominance of the limestone is everywhere obvious. It dominates, for example, the rapids of the 'Iron Gates', where the Danube roars through the Carpathians between Romania and Yugoslavia (Figure 1.5).

The point has therefore now been made with three examples that it is usually carbonate facies that are so remarkably persistent in a lateral sense. It might be said that these represent no more than quiet, low-energy conditions on an extensive continental shelf (or shelves). This does not fit, however, with the reef limestones and it certainly does not fit with the other facies that show the same persistence. As we continue down the stratigraphical column we find examples in other kinds of sediment and in other kinds of quite high energy facies.

THE 'GERMANIC' TRIAS

Every student knows, or should know, the classic trinity of the Germanic Triassic:

> Keuper
> Muschelkalk
> Bunter

It is not generally emphasised in textbooks, however, how very widespread these sediments are. Thus a Briton can drive through the Betic Cordillera in southern Spain and instantly recognise the gypsiferous, red and green marls of the Keuper. In the Celtiberic Mountains of eastern Spain, the three-fold division of the Trias is as clear as in Germany and the cross-bedded, red 'Bundsandstein' (for example, along the Rio Cabriel, south-west of Teruel) is exactly like the road-cuttings near Bridgnorth in the English Midlands. What is more, it is also exactly like the cliff sections in the Isker gorge, north of Sofia and elsewhere in Bulgaria, at the other end of Europe (Figure 1.6).

The basal conglomerate in England is full of boulders of a distinctive purple, 'liver-coloured' and white quartzites that have been matched with the *Grès de May* and the *Grès Armoricain* right across the other side of the English Channel in Brittany (though

Figure 1.6 *Lower Triassic sandstone at Belogradchik, north-west Bulgaria (DVA)*

I regard with some scepticism the notion that the boulders here travelled so far). Along the Rio Cabriel in Spain, it is the same, but there the source quartzite outcrops immediately below. Near Belogradchik, in north-west Bulgaria, again the basal conglomerate is largely composed of exactly similar purple quartzite pebbles (resting on Permian breccias also like those of Midland England). Even if one postulates continent-wide uplift to produce the conglomerate in such widely separated places, it is very difficult to explain why the source rock is also so remarkably similar from one end of Europe to the other.

Again, we can go even farther afield. It is well known that, apart from its basalts, the Newark Supergroup of the eastern seaboard of the United States is exactly like the Trias of north-west Europe, and both are now known to have been largely deposited in fault-controlled basins. The similarities are almost laughable, even to the extent of the 'Building Stones' of the basal Keuper near Birmingham, England, being remarkably like the sandstone which provided the 'brownstone' houses of much of old New York City. If we go to the High Atlas of Morocco, we find even closer similarities, with basic intrusions and extrusions within the familiar red sandstone.

However, when I make these comparisons across the Atlantic, I can almost hear my readers saying: 'plate tectonics'. Obviously if we close up the ocean again, the resemblances would not be so startling. Very well, but then let us go right down to the southwest corner of the United States and look at the Moenkopi and associated formations of Arizona. The glorious colours of the Painted Desert are produced by the same sort of red and green 'marls' as we Europeans have in our Keuper. The road cuttings along Highway 40 in Arizona show red and green marls and thin sandstones with layers of gypsum, all of which would be perfectly at home along the banks of the River Severn (Figure 1.7). Similar sediments with evaporites are seen in the cores of salt-domes below the Cupido Limestone mentioned above in north-east Mexico. What is more, a recent stamp from Argentina showed a red sandstone pinnacle in the depression of Ischigualasto (the Valley of the Moon) which a visiting Chinese immediately identified as Triassic. I have since seen red Upper Triassic mudstones near Daye in southern China.

THE COAL MEASURES

Again continental drift may be held to account for the remarkable similarity of the Upper Carboniferous (Pennsylvanian) Coal

Figure 1.7 *Road-cutting in the Triassic Moenkopi Formation, Highway 40 near Holbrook, Arizona, USA (DVA)*

Measures on both sides of what one of the airlines now likes to call the 'Atlantic River'. As a non-specialist, I found it quite easy, with my scanty knowledge of the plants of the British Coal Measures, to identify most of the diverse flora of the famous Mazon Creek locality in Illinois. Perhaps if I had been more of an expert the differences would have been more apparent, but experts always tend to obscure the obvious.

Certainly there are differences, especially in the better development of marine sediments in the American Pennsylvanian, but these in a way have obscured the resemblances; for work in America has concentrated on the marine fossils, whereas in Europe we have usually been forced to fall back on the non-marine faunas and floras. It is now known, however, that as with the plants, the non-marine bivalves of the American Mid-West are very like those that extend from Ireland to Russia.

Whatever the vertical and lateral changes in the Coal Measures, we still have to account for a general facies development in Late Carboniferous times that extends in essentially the same form all the way from Texas to the Donetz coal basin, north of the Caspian Sea in the USSR. This amounts to some 170° of longitude, and closing up the Atlantic by a mere 40° does not really help all that much in explaining this remarkable phenomenon. One looks in vain for a similar geographical situation at the present day. The nearest approach I can think of is around the deltas of south-east Asia.

LOWER CARBONIFEROUS LIMESTONE

In Britain, the great limestone development of Early Carboniferous (Mississippian) age used to be called the 'Mountain Limestone' because it formed so much of our upland scenery. In the early days of mapping in the United States, geologists (no doubt with a Europe-oriented education) had no difficulties in tracing the familiar 'Coal Measures' and the 'Mountain Limestone' of western Europe from the Appalachians right across the Mid-West.

The 'Carrière Napoléon' in the Boulonnais region of northern France (Figure 1.8) looks exactly like the 'Empire State quarry' in Indiana (Figure 1.9). Both use circulating wires to cut smooth faces in the Early Carboniferous limestone, and whereas the first

Figure 1.8 *Lower Carboniferous limestone, Carrière Napoléon, near Marquise (Pas de Calais), France (DVA)*

was used to build the high monument to the Grande Armée that overlooks Boulogne (with Napoleon at the top firmly turning his back on England), the Indiana quarry produced the stone facings for the Empire State Building in New York.

All the physiographical features of the Mid-Western Mississippian are familiar to the man from the English Pennines or the Mendips. The Mammoth Cave of Kentucky is nothing more than a rather larger Americanised version of Wookey Hole in Somerset or the Dan-yr-Ogof caves in South Wales.

However, this is a case where the stratigraphical wood cannot be seen for the nomenclatural trees. Whereas the British, in their old-fashioned way, have usually stuck to the general term 'Carboniferous Limestone' to cover all the varied carbonate facies of this age, the Americans—for very good local reasons—have

Figure 1.9   *Mississippian limestone, Empire State Quarry, near Bloomington, Indiana, USA (DVA)*

allowed the proliferation of formation names to obscure the unity of the whole.

So the Early Carboniferous was again a time of very widespread carbonate deposition. Not only were limestones deposited in Europe as far south as Cantabria and right across the Mid-West, they also went a lot further. Thus in Arizona, the Redwall Limestone of this age forms the steepest cliff in the Grand Canyon (Figure 1.10), and the name refers to the red staining of the rock from the overlying Permo-Triassic red beds, just as in the Avon Gorge at Bristol, the topmost Carboniferous Limestone is reddened by the overlying Permo-Triassic deposits. Similarly, right up in the Canadian Rockies, the Mississippian Rundle Limestone forms an impressive escarpment, for example above the town of Banff in Alberta (Figure 1.11). In Alaska, it is the Lisburne Limestone, with very similar characters.

We can also trace the early Carboniferous limestones in the opposite direction into Asia. Thus in Kashmir, there is a thick limestone of this age, very like its British counterpart and with a similar faunal list. The persistence of fossils will be the subject of the next chapter, but a cautionary note should perhaps be sounded here, for the familiarity of the fossil names may merely

Figure 1.10   *Redwall Limestone (Mississippian) forming the most obvious cliff in the Grand Canyon, as seen from the Powell Memorial, Arizona, USA (DVA)*

reflect the fact that they were studied by British palaeontologists. I am told that it also occurs in Western Australia.

FRASNIAN REEFS

Still climbing down the column, the next great limestone development we meet is in the lower part of the Upper Devonian. This is the Frasnian Stage and presents us with what is, perhaps, the most remarkable example of all. This was the heyday of reefs built by rugose corals and stromatoporoids. In some areas they started earlier (in the Givetian); elsewhere they lasted on into the Famennian, but in the Frasnian Stage reefs and reef limestones (in their broadest sense) were experiencing their finest

Figure 1.11 *Rundle Limestone (Mississippian), forming the escarpment of Mount Rundle, above Banff, Alberta, Canada (DVA)*

hour. This is true, in a humble way, in the so-called type area of the Devonian in south-west England. It is true in the classic reefs of Belgium, northern France and south-west Germany. It is true in the beautiful karst country of Moravia in central Czechoslovakia. It is also true in southern Morocco, in the American Mid-West and in the Canadian Rockies, where the cavernous rocks of these reefs form the most important oil reservoirs and the chief source of wealth in the province of Alberta (Figure 1.12).

In Western Australia too, magnificent reefs of this age are developed, perhaps the best in the world, notably in the splendid sections of the Windjana Gorge (Figure 1.13).

THE OLD RED SANDSTONE

The other great facies of the Devonian is the continental red sandstone development which extends across the north of Europe from Ireland to the Russian Platform. It has also been described from eastern Canada with fish remains very like those of the classic Scottish sections. It can also be seen below the

Figure 1.12 *Upper Devonian limestone escarpment with reef developments in the lower (Frasnian) part, Chinaman's Leap, near Canmore, Alberta, Canada (DVA)*

Carboniferous Limestone mentioned earlier in Kashmir. There, not only does the fish fauna closely resemble that of the Middle Old Red Sandstone in Scotland, but the sediments themselves are said to be exactly like the Thurso Flagstone Group of Caithness. So it is not merely marine deposits that are so incredibly persistent about the earth's surface. It is also found, with similar fish fossils such as *Bothriolepis*, in Australia.

MID-SILURIAN LIMESTONES

The great time of carbonate deposition during the Silurian Period was what we would call Wenlock time in Britain. The escarpment of Wenlock Edge in Shropshire is formed by a massive limestone

Figure 1.13   Panorama of a late Devonian reef, showing the massive reef proper in the centre, fore-reef talus deposits on the left and flat-flying lagoonal deposits on the right. Windjana Gorge, Napier Range, Western Australia (photograph kindly provided by Dr P. E. Playford)

(more thinly bedded below) of Mid Silurian age. Minireefs, or 'ballstones', are developed in most outcrops of this Wenlock Limestone, though they nowhere approach the size of the reefs discussed earlier. Similar reef limestones extend up to Scandinavia, where they reach their finest development in the island of Gotland out in the Baltic.

On a much grander scale are the Niagaran limestones around the Great Lakes in North America, where the reefs reach tremendous proportions, such as the splendid Thornton Reef on the outskirts of Chicago. But in time terms, these limestones are very much of the same age as those in Europe and, as I have said elsewhere, 'the Niagara Falls are nothing more than the Niagara River falling over an escarpment of Wenlock Limestone'.

ARENIG QUARTZITES

In the Ordovician of my corner of the world, the most remarkable example of persistence of facies is that of the purple and white

Figure 1.14   *Lower Ordovician* Grès armoricain *forming headland near Camaret, Crozon Peninsula, Brittany, France (DVA)*

quartzites in the lower part of the System. Every British geology student knows about the 'liver-coloured' quartzite pebbles which are found in our Triassic conglomerates (referred to earlier) and which are said to have come all the way from the Ordovician *Grès Armoricain* and *Grès de May* of Brittany (Figure 1.14), even though this implies the transportation of pebbles up to 20 or 30 cm diameter for several hundred kilometres up to the English Midlands. Perhaps it is the distinctiveness of this particular anatomical colour that makes us forget the pure white quartzites that also occur at this level, even in our own Midlands (as the Stiperstones Quartzite of west Shropshire). But whether our pebbles come all the way from Brittany (as still seems possible) or from hidden or eroded local sources is almost irrelevant, for the outstanding fact about these quartzites is their persistence. From England they go across to Brittany, where they are seen in sea cliffs and along into Normandy where they form, for example, the great escarpment at Falaise on which stands William the Conqueror's castle. From there they go right down to northern Spain. In Cantabria, the barrier they formed (as the Barrios Quartzite) played a major role for centuries in the military history of the Iberian Peninsula. West from there they are seen on the north coast in the gentler province of Galicia. They appear again on the south and east sides of the Spanish Meseta and then on way down into Africa. There are massive quartzites of similar type in the Ordovician of other parts of the world, from Bulgaria to the Canadian Rockies, but in my ignorance I will not risk saying that they are of exactly the same age. What is more, the reassemblage of continents, as now envisaged, does not simplify the picture at all; it makes it far more complex.

THE BASAL CAMBRIAN QUARTZITE

Even more remarkable than the basal Ordovician quartzite is the one that is found, almost all over the world, at the bottom of the Cambrian. Here dating becomes more and more problematical as the time spans become longer and longer. One is tempted to get mixed up with arguments about the origins of life and the beginning of the main fossil record, of the mysterious 'Lipalian Interval' that was once favoured, and of great world-wide marine transgressions. Perhaps all that it is safe to say in this context is that very commonly around the world one finds an unfossiliferous quartzite conformably below fossiliferous Lower Cambrian and unconformably above a great variety of Precambrian rocks. This is true wherever one sees the base of the Cambrian in Britain, it is true in east Greenland, it is true in the Canadian Rockies and it is true in South Australia. In fact it is even more remarkable than this, in that it is not only the quartzite, but the whole deepening succession that tends to turn up almost everywhere, i.e. a basal conglomerate, followed by the orthoquartzite, followed by glauconitic sandstones, followed by marine shales and thin limestones. In the northern Rockies one can even recognise at this level the 'Pipe Rock' of the Scottish Highlands—a bed full of borings known as *Skolithos*.

THE LATE PRECAMBRIAN GLACIATION

Finally, before we become completely lost in Precambrian fantasies, one must mention the glacial and periglacial deposits that occur in many parts of the world near the top of the known Precambrian. Dating has obviously become extremely inaccurate by this level compared with the precision higher up the stratigraphical column, but presumed Late Precambrian glacial deposits extend from the west of Ireland up through the Highlands of Scotland and then up through Norway to Varangerfjord near its northernmost tip. Very similar deposits are known in Greenland, which might be expected from drifting arguments, but they are also known in other parts of Europe and as far away as Brazil, China and Australia.

Like these, the 'red beds' of late Precambrian times have no accurate dates but are remarkably similar, whether they be the 'Torridonian' of Britain, the 'Huronian' of North America, the 'Vindhyan' of India or a dozen other named groups in all parts of the world, from China to Australia, Africa and South America.

These final examples of persistent facies bring us right up against the obvious explanation of a climatic control, and certainly strong arguments have been put forward on climatic grounds to explain some of the carbonate distributions. However, the object of this chapter is not to explain, but to wonder; the conclusions must come later.

So as one goes down the stratigraphical column, if one leaves behind the spectacles of the specialist and looks about one with the wondering eyes of a child, one never ceases to be amazed at the diversity and yet the uniformity of it all. No doubt my readers will have bigger and better examples of the persistence of facies which so fascinates me in this chapter, but I write as far as possible from my own experience. Someone will probably tell me of the almost incredible persistence of the Karroo sandstones of southern and eastern Africa, or the Nubian sandstones across northern Africa into the Middle East, or the Dakota Sandstone in the American West. I shall be glad to have my prejudices confirmed.

Before leaving this matter of persistence I want to mention a few special cases of persistence on a finer scale.

SPECIAL CASES

We have so far considered a number of examples of persistent facies on a rather grand scale; but there are more detailed examples of it which in their way are even more amazing.

Thus the Englishman, familiar with his Rhaetian sediments at the top of the Trias, cannot but be astounded when he reads of the Rhaetian deposits of Thailand and finds them described as black pyritous shales with *Rhaetavicula contorta*, resting on red marls and sandstones, with evaporites, just like those of the Severn cliffs. Certainly the Rhaetian I have seen in southern Europe, for example on the south side of the Pyrenees and along the shores of Lake Iseo in northern Italy, is incredibly like that of Britain, even in the way the fossils are preserved.

One of the most famous of all fossiliferous deposits is the Solenhofen Lithographic Stone of Bavaria in southern Germany. This fine-grained limestone is only developed over a very small area, usually thought of as a lagoonal deposit behind sponge reefs, though recently interpreted as deposits in offshore sediment traps.

Besides its famous specimens of *Archaeopteryx*, which have won so much popular attention, it has a marvellously preserved fauna and flora of great diversity, including ammonites which date it quite precisely as early Kimmeridgian in age.

This is wonderful enough; but in the western part of the southern French Jura, the Cerin Lithographic Stone is of similarly limited extent behind reefs, of similar lithology, of similar fauna and flora and of similar age (Figure 1.15). Unfortunately no *Archaeopteryx* has yet been found there, but otherwise the resemblance is startling.

Curiouser and curiouser, on the other side of the Pyrenees in Spain, in the high escarpment near the attractive village of Ager, north of Lerida, is yet a third lithographic limestone, again restricted to a very small area and exactly like the other two in lithology, fauna, flora and age (and even yielding a feather). Other similar deposits of about the same age are said to occur

Figure 1.15 *Cerin Lithographic Stone, Upper Jurassic (Kimmeridgian) near Cerin (Ain), French Jura (DVA)*

at Nusplingen in the Swabian Jura, at Talbrager in New South Wales and in the central Congo.

The above are selected examples from among many, of discontinuous distributions of similar synchronous deposits. There are even more examples of very thin units that persist over fantastically large areas in particular sedimentary basins. Lithological units of 30 m or less in the Permian of western Canada, have been shown to persist over areas up to 470 000 km². The thin basal member of the Trias, about a metre thick, can be found all round the Alpine chain.

The occurrence of banded ironstones around the world in late Precambrian rocks is well known. Particularly noteworthy is the economically important Animikie Basin, with the fabulous Mesabi, Marquette and other ranges, at the west end of Lake Superior in North America. Others of about the same age (i.e. about 2000 million years BP) are the Transvaal Basin in South Africa, the Hamersley Basin in Western Australia and the Dharwars Series of India. All have the banded or varved iron formations that are characteristic of this episode in earth history. Even more remarkable, however, is the fact that individual bands can be traced over vast areas. Thus in the valuable Brockman Iron Formation of the Hamersley Basin, bands 2–3 centimetres thick are said to be correlatable over an area of some 52 000 km² and even microscopic varves within those bands can be traced over nearly 300 km.

Like wartime bomb stories, every geologist seems to have his own favourite example to cap all others. Like a politician, I may have overstated my case to make my point. Those who are fascinated by the minutiae of stratigraphical correlation may be horrified at my generalisations. But I find myself left with what may be called the first main proposition of this book:

At certain times in earth history, particular types of sedimentary environment were prevalent over vast areas of the earth's surface. This may be called the 'Phenomenon of the Persistence of Facies'.

REFERENCES

Ager, D. V. (1958). 'On some Turkish Sediments', *Geol. Mag.*, Vol. 95, pp. 83–84.
A brief note recording (among other things) north-west European type Chalk on the south side of the Black Sea.

Ager, D. V. (1968). 'The Lateral Persistence of Carbonate Facies in Europe and Adjacent Areas, Rep. 1968 Nat. Conf. Earth Sci., Banff, Alberta, pp. 16–20.

A duplicated report summarising the main points of this chapter.

Ager, D. V. (1970). 'On Seeing the Most Rocks', *Proc. Geol. Ass.* (H. H. Read volume), Vol. 81, pp. 421–427.

A light-hearted contribution to a *Festschrift* drawing attention to some of the points made in this chapter.

Ager, D. V. (1980). *The Geology of Europe.* McGraw-Hill, Maidenhead, 535 pp.

In which I recklessly try to summarise the geology of our fascinating continent and draw attention to many of the examples of the persistence of facies discussed in this chapter.

Andrichuk, J. M. (1958). 'Stratigraphy and Facies Analysis of Upper Devonian Reefs in Leduc, Stettler and Redwater Areas, Alberta', *Bull. Am. Ass. Pet. Geol.*, Vol. 42, pp. 1–93.

Classic paper on the late Devonian reefs of western Canada.

Curry, D., Gray, F., Hamilton, D. and Smith, A. J. (1967). 'Upper Chalk from the Sea Bed, South of Cork, Eire', *Proc. Geol. Soc., London*, No. 1640, pp. 134–136.

A further westerly extension of Upper Cretaceous Chalk.

Dehm, R. (1956). 'Zeitgebundene Gesteine und Organische Entwicklung', *Geol. Rundschau*, Vol. 45, pp. 52–56.

A brief paper on the persistence of the late Precambrian iron ores and the late Jurassic lithographic limestones.

Gupta, V. J. and Dension, R. H. (1966). 'Devonian Fishes from Kashmir, India', *Nature, Lond.*, Vol. 211, pp. 177–178.

On the Old Red Sandstone facies in Kashmir.

Lecompte, M. (1958). 'Les Récifs Paléozoiques en Belgique', *Geol. Rundschau*, Vol. 47, pp. 384–401.

It is difficult to select one from Lecompte's voluminous works on the Devonian reefs of Belgium, but this gives the general picture.

Lowenstam, H. (1957). 'Niagaran Reefs in the Great Lakes Area', *Geol. Soc. Amer.*, Vol. 67, pp. 215–248.

One of a series of notable papers on these mid-Silurian reefs.

McGugan, A. (1965). 'Occurrence and Persistence of Thin Shelf Deposits of Uniform Lithology', *Bull. Geol. Soc. Amer.*, Vol. 76, pp. 125–130 and 615–616.

On persistent units in the Canadian Permian.

Nikolov, I. (1969). 'Le Cretacé Inférieur en Bulgarie', *Bull. Soc. Géol. France*, Vol. 11, pp. 56–68.

Nikolov, I. (1971). 'Über die Lithofazies der Barreme-Ablagerungen in Bulgarien', *N. Jb. Geol, Palaönt. Abh.*, Vol. 139, pp. 163–168.

Two accounts of the Urgonian and associated rocks at the other end of Europe, showing the lenticular nature of this 'reef' facies.

Playford, P. E. and Lowry, D. C. (1967). 'Devonian Reef Complexes of the Canning Basin, Western Australia', *Bull. Geol. Surv. W. Australia*, Vol. 118.

Definitive paper on these remarkable late Devonian Australian reefs.

Spjeldnaes, N. (1967). 'The Palaeogeography of the Tethyan Region During the Ordovician', in C. G. Adams and D. V. Ager (eds) *Aspects of Tethyan Biogeography*, Systemics Ass. Publ., No. 7, pp. 45–57.

Discusses the persistence of certain Ordovician facies.

Trendall, A. F. (1968). 'Three Great Basins of Precambrian Banded Iron Formation Deposition: A Systematic Comparison', *Bull. Geol. Soc. Amer.*, Vol. 79, pp. 1527–1544.

A general comparison of the West Australian, South African and North American basins.

Trueman, A. E. (1946). 'Stratigraphical Problems in the Coal Measures of Europe and North America', *Quart. J. Geol. Soc., Lond.*, Vol. 102, pp. 49–86.

Classic paper on the Late Carboniferous of the northern hemisphere.

Van Straaten, L. M. J. U. (1971). 'Origin of Solnhofen Limestone', *Geol. Mijnbouw*, Vol. 50, pp. 3–8 (1971).

Proposes offshore sediment traps rather than back-reef lagoons.

Wallace, P. (1972). 'The Geology of the Palaeozoic Rocks of the Cantabrian Cordillera, North Spain', *Proc. Geol. Ass.*, Vol. 83, pp. 57–73.

A general account of this wonderful Palaeozoic area, drawing attention to some of the persistent facies.

Walsh, P. T. (1965). 'Cretaceous Outliers in South-West Ireland and Their Implications for Cretaceous Palaeogeography', *Proc. Geol. Soc., Lond.*, Vol. 1629, pp. 8–10.

Records a remarkable occurrence of Late Cretaceous Chalk in south-west Ireland.

2
The Fleeting Fossil

We all know that species come and go with frightening rapidity (in fact *Homo sapiens* has already exceeded the life expectancy of most species). We also know that the fossil record is fragmentary in the extreme. Yet it is the common experience of most palaeontologists that, just as lithological facies are persistent around the world, so are the fossils which they contain. Theoretically, we might expect this to be so, since the same environment tends to support the same kind of organisms, but in fact the persistence of some fossils appears to go far beyond what we know at the present day.

I have already written about the geographical distribution of fossils (especially Mesozoic brachiopods) at nauseating length, so it is not appropriate to do so again. But I cannot resist mentioning two or three examples which are very well documented and which I believe in personally (one gets very sceptical about other people's records in palaeontology).

One example is the late Triassic brachiopod *Halorella*, which is large and distinctive and which has been recorded from every continent except Antarctica, in rocks dating from quite a small span of late Triassic time (Figure 2.1a). It has no apparent direct ancestors or descendants, yet it turns up simultaneously in places as far apart as Indonesia, Siberia, Turkey, New Zealand and Nevada. What is more, these forms are not only the same genus, but also the same species, and the Nevadan specimens have even (very justifiably) been put in the same subspecies as a form in the Austrian Alps.

A distant relation of *Halorella*, called *Peregrinella*, is even more remarkable in early Cretaceous rocks (Figure 2.1b). It is best known from the presbytery garden at Chatillon-en-Diois in the French Alps, but has also been found in a single block in Poland, as a single specimen in Czechoslovakia, in southern China, in California and at not more than two or three other places in the world. Yet it is one of the most distinctive brachiopods in the whole record and it has internal structures which make it clear that none of the abundant brachiopods in the strata above or below could possibly be classified as even distant relations. Its name means, in fact, 'little stranger', though it is by no means small for a brachiopod.

In other words, we have fossils that just suddenly appear around the world at one moment in geological history and 'whence, and whither flown again, who knows'? One can understand this, perhaps, in the fragmentary record of a rare and little-known group, but the Mesozoic brachiopods are now very thoroughly documented in every stage and the relations of these large and distinctive forms can hardly have been missed.

I use Mesozoic brachiopod examples because they are the only ones I can believe in wholeheartedly from my own work, but probably every specialist can produce his own examples. At the other end of the animal kingdom, I might cite that well-loved dinosaur *Iguanodon*, now identified in Africa, Australia, Asia and Spitzbergen, besides its best-known English and Belgian haunts.

Even taking into consideration phyla or classes as a whole, one is struck (at least I am struck) by the remarkable way in which particular groups of fossils seem to have been 'in fashion' for a while and then return to a comparatively minor role. Why was the Cambrian 'the Age of the Trilobites' as the popular science books tell us? We know that there were plenty of other animal groups around, but in that period the trilobites seem to have dominated almost everywhere. Similarly, the brachiopods and corals in the early Carboniferous, the bivalves in the early Jurassic and the echinoids in the Miocene.

I sometimes wonder at the sheer multiplicity of the oyster *Gryphaea arcuata* there must have been on the earliest Jurassic sea-floors over what is now Europe, and the millions upon millions of creatures (whatever they were) that chewed their way through Jurassic mud to produce the trace-fossil *Zoophycos*.

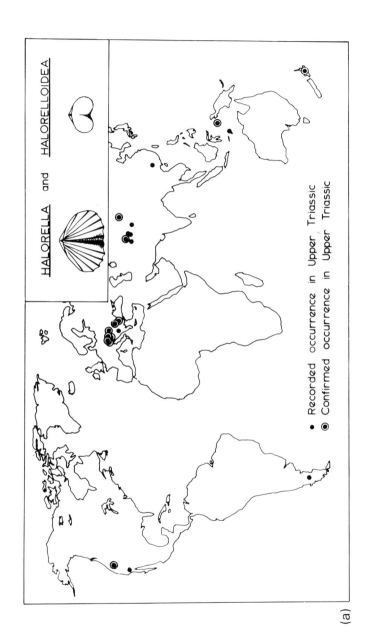

Figure 2.1 (a) Distribution of the distinctive brachiopods Halorella *and its close relation* Halorelloidea *in the late Triassic*

(a)

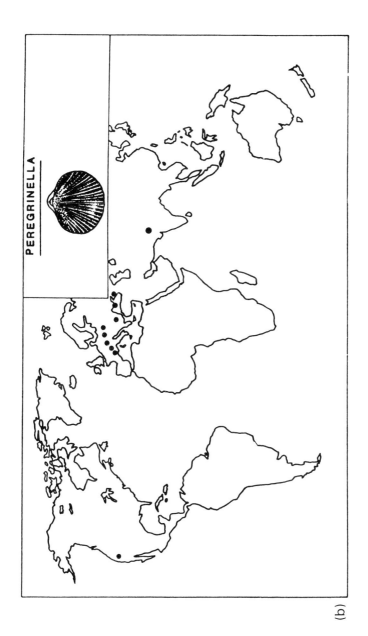

Figure 2.1 (b) Distribution of Peregrinella in the early Cretaceous

(b)

One can easily produce evidence at the present day of great local abundance (e.g. of starfish or pilchards), but I know of nothing on a modern sea-floor to compare with the abundance *plus* wide distribution of the examples just mentioned.

It may be said that these are superficial impressions derived from a superficial knowledge of small parts of the stratigraphical record. But I defy anyone looking at any part of the earth's history closely and with an open mind, not to be struck by the super-abundance and ubiquity of particular fossils. I know very well that in many cases this is in the mind of the palaeontologist rather than in the rocky facts themselves. Certain fossils have, for some inexplicable reason, taken root in the literature and palaeontologists' subconscious and so been recorded every-where. *Atrypa reticularis* in the Silurian and *Stringocephalus burtini* in the Devonian are obvious examples.

There is much subjectivity in systematic palaeontology. I have pointed out in an earlier essay, for example, that the distribution of certain fossils corresponds remarkably well with the old Austro-Hungarian Empire, though it is doubtful whether the Hapsburgs had much control over evolution. The old sunset-defying British Empire stands out even better, and it so happens that the main fossiliferous parts of it fall on a non-equatorial great circle, thereby providing ample ammunition for the polar wanderers. But clearly it was imperialistic palaeontologists rather than imperialist fossils that set the pattern in both cases.

Nevertheless, even allowing for all the frailties of palaeontolo-gists, there still remains a remarkable picture of palaeontological persistence. The Late Carboniferous forests around the northern hemisphere were far more alike than the modern ones. The mere fact that all the classic ammonite zones of the European Lower Jurassic have been recognised, in the right order, in North America is surely remarkable, as is the persistence of practically all the Ordovician and Silurian graptolite zones from the UK to China. Even accepting all the presumptions of stratigraphical palaeontology, there still remains the fact that the zonal fossils are not merely successive segments of a branching tree of evolution. In most cases they are segments of different branches, and very often they are different trees, that come and go in their relative importance. Thus all 11 of the zonal ammonites of the Lower Lias (basal Jurassic) belong to different genera and only two of them even belong to the same sub-family. And while the fashionable ammonites and graptolites were competing for their place in the spotlight, the minor characters in the stratigraphical play were doing exactly the same thing. For example, pretty well all over the world, the Lower Jurassic terebratulids were represented by the one monotonous, dull-looking genus *Lobothyris*, though the order was more complex both before and afterwards, and several other families somehow survived this temporary eclipse. The crinoids, the belemnites and the corals behaved similarly.

Pyramids of paper have been piled up on the subject of faunal provinces in the fossil record, but very few of them stand up to critical examination and even the latest symposia on the subject have produced very little that can be regarded as concrete evidence. Almost all the differences that have been noted in contemporaneous fossil faunas and floras can be explained in terms of local environmental differences that are reflected in the sediments. Thus the famous 'Bohemian' (or 'Hercynian') and 'Rhenish' provinces of Devonian times are little more than the differences between a lime-mud and a sandy sea-floor.

The only distinctive 'province', in the strict sense, that I have recognised in my studies of Mesozoic brachiopods, is a somewhat vaguely defined equatorial belt. Thus in the Bathonian stage of the Middle Jurassic, the distinctive genus *Flabellothyris* turns up along a belt parallel to the present equator (for example, in Mexico, Morocco and India) and not to the north or south. But this still confirms my point, since such forms spring up and disappear as if by the act of an experimenting (and frequently unsuccessful) creator.

Many of the sort of examples I have cited so far have been pushed aside by theorising palaeontologists with the disparaging term 'facies fossils'. In other words they appear and disappear as they do because they are controlled by a particular sedimentary facies. Certainly the only reasonable explanation for some of these occurrences (such as those of *Peregrinella*) seems to be that the organisms concerned lived in very restricted environments that did not normally get preserved, though this hardly explains their sudden appearance without obvious ancestors, with a wide but discontinuous distribution. In other words, they appear at the same time at widely separated localities. What is more, it cannot be emphasised too strongly that *all* fossils are 'facies fossils'.

If there were nothing more to it than this, I would only be repeating what I said in the previous chapter about the persistence of facies. But surely some fuller explanation is needed. Surely somewhere in the world there would have been the right facies preserved to provide the immediate ancestors of *Peregrinella* or *Iguanodon*? Surely somewhere in the world there should be the right facies preserved to provide the connecting links between the Palaeozoic faunas of the late Permian and the Mesozoic faunas of the mid and late Trias? I know that enthusiastic palaeontologists in several countries have claimed pieces of this missing record, but the claims have all been disputed and in any case do not provide real connections. That brings me to the second most surprising feature of the fossil record. Alongside the theme of the geographical persistence of particular fossils, we have as a corollary, the abruptness of some of the major changes in the history of life.

It is both easy and tempting (and very much in line with the other ideas expressed in this book) to adopt a neocatastrophist attitude to the fossil record. Several very eminent living palaeontologists frequently emphasise the abruptness of some of the major changes that have occurred, and seek for an external cause. This is a heady wine and has intoxicated palaeontologists since the days when they could blame it all on Noah's flood. In fact, books are still being published by the lunatic fringe with the same explanation. In case this book should be read by some fundamentalist searching for straws to prop up his prejudices, let me state categorically that all my experience (such as it is) has led me to an unqualified acceptance of evolution by natural selection as a sufficient explanation for what I have seen in the fossil record. I find divine creation, or several such creations, a completely unnecessary hypothesis. Nevertheless this is not to deny that there are some very curious features about the fossil record.

One thing which has struck me very forcibly through the years is that most of the classic evolutionary lineages of my student days, such as *Ostrea–Gryphaea* and *Zaphrentis delanouei*, have long since lost their scientific respectability, and in spite of the plethora of palaeontological information we now have available, there seems to be little to put in their place. In 40 years' work on the Mesozoic Brachiopoda, I have found plenty of relationships, but few if any evolving lineages. This does not mean that I deny evolution occurred. It would be even more difficult to explain my data if it did not. What it seems to mean is that evolution did not normally proceed by a process of gradual change of one species into another over long periods of time. I have long criticised the notion that evolution can be studied by chasing fossil oysters up a single cliff, even though that approach was first brilliantly expounded by my most distinguished predecessor at Swansea. One must clearly study the variation of a species throughout its geographical range, at one moment in geological time, before one can claim that it has changed into something else.

I am now very much of the opinion that most evolution proceeds by sudden short steps or *quanta* as brilliantly expressed in recent years by Stephen Gould and Nils Eldredge, who suggested that new species arise, not in the main centre of its ancestors, but in peripheral, somewhat isolated populations. As Gould expressed it: 'The history of evolution is not one of stately unfolding, but a story of homeostatic equilibria, rarely disturbed by rapid events of speciation.' He therefore concluded that many of the awkward breaks in the fossil record are more real than apparent. So we come to a somewhat catastrophic attitude to evolution. It is also probable, of course, that the peripheral populations will encounter the environmental differences that will select particular characters.

The greatest problems in the fossil record, however, are the sudden extinctions. Examples such as the disappearance of the dinosaurs have been chewed over and over *ad nauseam*, with every possible cause blamed, from meteoric impact to chronic constipation. For any one ecological group, such as the dinosaurs, it is comparatively easy to find a possible cause. It is much less easy when one has to explain the simultaneous extinction of several unrelated groups, ranging from ammonites to ptero-dactyls, living in different habitats at the end of the Mesozoic. This is not to say that we have to fall back on Old Testament catastrophism. At the opposite end of the scale from the fundamentalists, we have such interesting hypotheses as the one put forward recently drawing attention to the close positive correlation between the susceptibility of groups to extinction and the oxygen consumption level of their modern representatives.

Again I must say that I am very aware of the frailties of palaeontologists. It is very easy to demonstrate a break between Palaeozoic life and Mesozoic life if Palaeozoic specialists only work on Palaeozoic fossils and if Mesozoic specialists only work on their Mesozoic descendants. This was impressed on me very strongly as one of R. C. Moore's regiment of workers producing Volume H of his vast *Treatise on Invertebrate Paleontology*. Not only did the names change at the Palaeozoic/Mesozoic boundary, but also the classification, the methods of study and even the terminology of the anatomical parts. Our great editor was moved, as a result, to insert a sentence or two commenting on the major changes at this level in my particular group. I had the temerity to object to these insertions, because I knew full well from years of grubbing in the darker recesses of the phylum that the changes were in the mind rather than in the matter. It has also become evident from recent literature that there was nothing like a 'simultaneous' extinction of many different groups, either within the brachiopods alone or within the organic world in general, at the end of the Permian; I am told that plant spores, at least, still show an uncannily rapid change at this level all round the world, though the big change in plant macrofossils seems to have come much later.

Nevertheless, great changes do occur and have been well documented. There is no point in denying them, even though we may justifiably quibble about the subjectivity of palaeontologists, the imprecision of the boundaries and the importance of the hidden gaps. Certainly there are not enough of such sudden changes to build a stratigraphy. Thus of the two most notable recent protagonists of major faunal breaks, the great German palaeontologist Professor Schindewolf claimed three, and the eminent American palaeontologist Professor Newell claims six (Figure 2.2). To each of these may be added another if we count the faunal extinctions brought about by man in the last million years or so. This last, however, has hardly affected the marine invertebrate faunas on which so much of the rest of our record depends.

These major breaks are all, of course, like the Permian/Triassic break just discussed, the limits between Periods and Eras. It is a

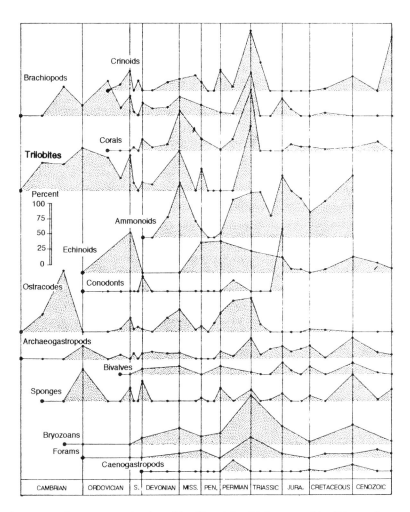

Figure 2.2 *Percentage extinction of families of the chief groups of fossil invertebrates. Notice the correlation of peaks near the end of the Devonian, Permian, Triassic and Cretaceous periods (from Newell (1967) by kind permission of Professor Newell and the Geological Society of America)*

curious fact, however, that some of the most abrupt and startling changes in the fossil record are the least obvious. I have long been impressed, for example, by the changes that occur between the Frasnian and Famennian stages of the Late Devonian. Thus in the reefs of Windjana Gorge (mentioned above) the stromatoporoids are suddenly replaced by calcareous algae. The point which has struck me most (in my myopic way) has been the abrupt disappearance at this level, all over the world, of the atrypid brachiopods. I have suggested that they were replaced ecologically by the rhynchonellids. It has since been pointed out by Digby McLaren (1970) that many other groups disappeared at the same level or underwent traumatic changes. Thus the corals and the stromatoporoids virtually ceased abruptly, though presumably some must have survived somewhere to produce the later faunas of Carboniferous times. Besides these major groups, almost all the trilobites, the tentaculitids and several other brachiopod groups (the orthids, pentamerids and stropheo-dontids) went the same way. After this calamity, several new groups appeared, of which the most important were the clymeniid ammonoids, quite different from anything that went before and destined to play a major role in the world of stratigraphical palaeontology.

McLaren's suggested explanation for this particular catastrophe arose from our recent knowledge (from Mars, Venus, Mercury and the back of the moon) that meteorites and meteoritic impacts are a general feature of the Solar System. There are now well over a hundred such impact structures claimed on earth. Many of these are, and will be, disputed, but it still seems probable that such collisions must be considered as a regular feature of earth history. It has been suggested that a giant meteorite falling in the Atlantic would produce a wave 6000 m high. As McLaren says: 'This will do'. Such a wave and the turbidity that followed would kill off all the bottom-dwelling filter feeders around the world, except perhaps those in some sheltered inland seas. Then, when the wave and the turbidity subsided, the shallow water would have been repopulated from other habitats.

Since writing the above in the first and second editions of this book, extra-terrestrial impacts have become a growth industry. Several other authors have suggested them as a major feature in earth history, notably Urey (1973) and Clube (1978), but the greatest interest was aroused by the work of Alvarez, father and son, together with colleagues (1979, 1980) who found a high level of iridium at the Cretaceous–Tertiary boundary near Gubbio in the Italian Apennines. Iridium is an element that is characteristic of meteorites and is not normally found on the surface of the earth. It is still only in very minute quantities, but has turned up at the same level at many other localities as widely scattered as Denmark, Spain and New Zealand. It was therefore postulated that a large body struck the earth at this moment in time and caused widespread extinctions, notably that of the dinosaurs, which so much dominate the popular imagination. Similar anomalies have been found at other critical levels such as the top of the Palaeozoic. Perhaps the most persuasive work on the subject was that by Hsü (1986). Naturally such ideas appeal to the catastrophist in me, but it is only fair to state that there has been opposition to the idea, both in respect of the origin of the iridium and to the likelihood of the impacts having caused mass extinctions. Reverting as always to the principle of what I have seen for myself, I must express healthy scepticism. I have discussed all this at greater length in another book (Ager, in press).

One of the remarkable events recorded in the geological record is the nuclear chain reaction that seems to have occurred in the vicinity of the Oklo uranium mine in Gabon some 1800 million years ago over a period of about half a million years.

No doubt many of my readers will instantly distrust such apocalyptic explanations as meteoritic impact, nuclear explosion or equally the increase in cosmic ray activity favoured by other palaeontologists, such as Schindewolf (1962). For a time it was popular to suggest that reversals in the earth's magnetic field, which we know to have been sudden, may have temporarily broken down the protective shield provided by the van Allen Belt against cosmic rays and so stimulated evolution by way of genetic mutation. However, it is now argued that the increased radiation would be no more than the existent difference between the poles and the equator. Nevertheless, whatever the mechanism involved, there is an increasing volume of evidence of a strong correlation between extinctions and magnetic reversals. Thus six out of eight species of radiolarians that have become extinct in the last 2½ million years show a very strong correlation between their last appearance and a reversal horizon. This has been observed in a large number of deep sea cores from both high and low latitudes, though one is tempted to ask how many other species did not become extinct during this period. Naturally, if one goes farther back in time the evidence becomes more shaky, but nevertheless there apparently were reversals in late Permian and late Cretaceous times which coincide with the two most important extinction levels.

In any case, there are earthier explanations, such as the changes in sea-level favoured by Newell (1963, 1967), or changes in climate (perhaps themselves brought about by changes in the earth's magnetism) which are favoured by many authors. We shall hear more of this later. Perhaps it is better, rather than roaming back into the recesses of geological time to consider the sudden and simultaneous extinctions that happened in our geological yesterday. These did not result from the Pleistocene glaciation (in fact none of the ice ages of the past appear to have coincided with mass extinctions), but all occurred after the final retreat of the ice. Between 7700 and 8000 years ago, within three short centuries, the Columbian mammoth, the dire wolf, the camel, the horse, the giant armadillo and the western bison all became extinct in North America. Many other species had disappeared within the two preceding millennia. This is commonly blamed on a prolonged drought during the amelioration of climate following the last glaciation. It is also blamed on the human technologies of slaughter and environment destruction. There is no doubt that *Homo sapiens* has been one of the most efficient agents of extinction in the history of the earth. He may already have been responsible for the extinction of some 450 species. It has been said, for example, that of all the species alive in Olduvai Gorge in Tanzania at the time early man arrived there, only 60% survived to the present day. But, after all, man is just one more species in the long history of the earth.

We have seen very recently in the South Pacific how one species can rapidly wipe out vast populations of other organisms. In a few years, the Crown of Thorns starfish, *Acanthaster planci*, is said to have decimated the corals of the Great Barrier Reef and of many island groups. This in turn may have been caused by the fact that the gastropod that previously controlled the echinoderm was prized by human tourists. More recently, more prosaic workers have claimed that the damage to coral reefs is nowhere near as widespread as the earlier alarmist reports had suggested. The situation is probably a cyclic one anyway, with man only partly responsible for the present coral massacre.

Nevertheless, it has been said that virtually all the extinctions of the last million years or so can be blamed, with fair confidence, on the intelligence (or lack of it) of *Homo sapiens*. We could say (as I have said myself) that we are just one more new species, albeit more destructive than most, having its effect upon its contemporaries. But clearly, we cannot blame a single organic agent for the simultaneous extinction of all the varied and unrelated groups of dinosaurs, the pterosaurs, the marine reptiles, the ammonites, the belemnites, the rudistids and many minor groups besides at the end of the Mesozoic. Certainly, if we exclude our own species, we cannot find any one factor having this sort of effect.

Almost all the theories (including the Noachian one) that seek to explain major extinctions in the past, lead by one route or another to climatic oscillations and related matters such as the composition of the earth's atmosphere. These in turn tend to point to extra-terrestrial phenomena. Hypotheses such as fluctuations in solar radiation come up again and again (recently, for example, on the evidence of coral reefs around Barbados). The Nobel Prize-winner, Professor H. C. Urey, came out strongly in support of extra-terrestrial causes for mass extinctions. He suggested that rare collisions between the earth and comets, recorded as scatters of tektites, must have produced vast quantities of energy that would have been sufficient to heat up considerably both the atmosphere and the surface layers of the oceans. The resultant high temperatures and high humidities could have had a disastrous effect on both land and marine faunas. We are always forced back on seeking some control outside and greater than the earth itself. This, of course, is easier than finding provable hypotheses for which we may expect evidence within the earth's crust. But in its way it is more exciting and more of a challenge than the current enthusiasm for plate tectonics and sea-floor spreading. After all, the physical facts of life are commonplace throughout the universe; the biological, so far as we know, are peculiar to this planet and then for only a very brief part of its history. Life has always been on a razor-edge of survival and it is surely important to understand those moments in the past when the organic world seemed closer than usual to obliteration.

It is essential to add that plate tectonics has crept into this matter as it has into every other aspect of the earth sciences. In particular, three major faunal boundaries can now be related to the plate situation in the North Atlantic area.

Since the first edition of this book my own views have inclined more and more towards major transgressions and regressions as the main cause of organic evolution and extinction. These may also be responsible for the associated climatic phenomena and may be blamed on the effects of plate tectonics with the rise and fall of mid-ocean ridges and the displacement of water by the piling up and erosion of new orogenic belts.

We cannot demonstrate anything really comparable to the sudden mass extinctions of the past happening at the present day, and in the fleeting second we have available that is hardly surprising. What is more, we cannot even see the processes going on today that might lead to such extinctions. I feel that we rely too much on the present state of affairs, too much on uniformitarianism, when interpreting the fossil record, especially in those groups that are now completely extinct or but a shadow of their former selves. It may be said of many palaeontologists, as Professor Hugh Trevor-Roper said recently of eighteenth-century historians: 'Their most serious error was to measure the past by the present.' We may arrive, therefore, at the second proposition of this book: *palaeontologists cannot live by uniformitarianism alone*. This may be termed the 'Phenomenon of the Fallibility of the Fossil Record'.

REFERENCES

Ager, D. V. (1968*a*). 'The Famennian Takeover', *Circ. Palaeont. Ass.*, No. 54a.
    Duplicated summary of a lecture on the abrupt faunal changes at the Frasnian-Fammenian boundary.
Ager, D. V. (1968*b*). 'The Supposedly Ubiquitous Tethyan Brachiopod *Halorella* and Its Relations', *J. Palaeont. Soc. India*, Vol. 5–9 (for 1960–1964), pp. 54–70.
    Records the remarkable distribution of *Halorella*, *Pereginella* and their relations.
Ager, D. V. (1976). The Nature of the Fossil Record', *Proc. Geol. Ass.*, *Lond.*, Vol. 87, pp. 131–160.
    A presidential address intended as a companion to this book, briefly expounding the author's palaeontological philosophy.
Ager, D. V. (in press). 'The New Catastrophism.' Cambrige University Press.
Alvarez, L. W., Alvarez, W., Asaro, F. and Michell, H. V. (1979). 'Anomalous Iridium Levels at the Cretaceous/Tertiary Boundary at

Gubbio, Italy: Negative Results of a Test for a Supernova Origin', in W. K. Christensen and T. Birkelund (eds), *Cretaceous–Tertiary Boundary Events*, Vol. 2, pp. 50–53.
The important paper in which attention was first drawn to the iridium anomalies.

Alvarez, L. W., Alvarez, W., Asaro, F. and Michell, H. V. (1980). 'Extra-terrestrial cause for the Cretaceous–Tertiary extinction', *Science*, Vol. 208, pp. 1095–1108.
Where the Cretaceous–Tertiary extinction is all blamed on a cosmic collision.

Clube, S. V. M. (1978) 'Does Our Galaxy Have a Violent History?', *Vistas in Astronomy*, Vol. 22, pp. 77.
The first of a series of works in which cosmic collisions are postulated.

Copper, P. (1967). 'The *Atrypa zonata* Brachiopod Group in the Eifel, Germany', *Senckenberg. Lethaea*, Vol. 47, pp. 1–55.
Records the abrupt extinction of atrypid brachiopods at the end of Frasnian times.

Gould, S. J. (1971). 'Speciation and Punctuated Equilibria: An Alternative to Phyletic Gradualism', *Abstracts 1971 Annual Meeting Geol. Soc. Amer.*, pp. 584–585.
Only the abstract of an exciting paper.

Gould, S. J. and Eldredge, N. (1977). 'Punctuated Equilibria: The Tempo and Mode of Evolution Reconsidered', *Paleobiology*, Vol. 3, pp. 115–151.
A further look at the episodic nature of evolution, summarising the real evidence very succinctly.

Hsü, K. (1986). *'The Great Dying.'* Harcourt Brace Jovanovich. San Diego, New York, London. 292 pp.

McAlester, A. L. (1970). 'Animal Extinctions, Oxygen Consumption, and Atmospheric History', *J. Paleont.*, Vol. 44, pp. 405–409.
For its interesting correlation of oxygen consumption and susceptibility to extinction.

McLaren, D. J. (1970). 'Presidential Address: Time, Life and Boundaries', *J. Paleont.*, Vol. 44, pp. 801–815.
Discusses (among other things) the multiple extinctions at the end of the Frasnian.

Middlemiss, F. A., Rawson, P. F. and Newall, G. (eds) (1971). 'Faunal Provinces in Space and Time', *Geol. J. Spec. Issue*, No. 4.
Twelve papers (including one by the author) and an excellent summing-up on this vexed question, with remarkably divergent views.

Newell, N. D. (1963). 'Crises in the History of Life', *Scient. Am.*, Vol. 208, pp. 77–92.
A semi-popular account of the periodic mass extinctions recorded in the fossil record.

Newell, N. D. (1967). 'Revolutions in the History of Life', *Spec. Paper Geol. Soc. Amer.*, No. 89, pp. 63–91.
A more detailed discussion of the same matter with some very useful figures.

Rhodes, F. H. T. (1967). Permo-Triassic Extinctions in 'The Fossil Record'. *Spec. Publ. Geol. Soc., Lond.*, pp. 57–76.
A paper confirming this author's own prejudices that the supposed great changes at the end of the Palaeozoic are more apparent than real.

Schindewolf, O. H. (1962). 'Neokatastrophismus', *Deutsche Geol. Gesell, Zeitschr.*, Vol. 114, pp. 430–445.
The views of this great German palaeontologist on major breaks in the fossil record.

Urey, H. C. (1973). 'Cometary Collisions and Geological Periods', *Nature, Lond.*, Vol. 242, pp. 32–33 (1973).
A very distinguished physicist expressing unconventionally catastrophic ideas about such matters as the extinction of the dinosaurs and citing evidence from tektites of extra-terrestrial impacts.

Vogt, P. R., Johnson, G. L., Holcombe, T. L., Gilg, J. G. and Avery, O. E. (1971). Episodes of sea-floor spreading recorded by the North Atlantic basement. *Tectonophys.*, Vol. 12, pp. 211–234.

3
More Gaps than Record

The late unlamented Field Marshal Goering once said that when he heard the word 'culture', he reached for his gun; I feel rather the same about the phrase 'continuous sedimentation'. What do we mean by 'continuous sedimentation'? Do we mean something like one sand grain every square metre of sea-floor per minute, per day, per year? Even the least of these would give us vastly more sediment than we normally seem to find preserved for us in our stratigraphical record. When attempts have been made to calculate rates of sedimentation in what look like continuously deposited sediments, the results look ridiculous. Thus the *Globigerina* ooze on the floor of the Indian Ocean seems to be accumulating at between 0.25 and 1 cm per 1000 years. A very conservative estimate for the Upper Cretaceous Chalk in northern Europe would give a figure of something over 9000 m as an absolute maximum, before consolidation, and about 30 million years for its deposition. That works out as around a third of a millimetre per year, or 200 years to bury an ammonite! And that is for the most rapidly accumulating chalk.

Recent estimates of the rate of deposition of deep-sea ooze show quite considerable variation, but still very little sediment. Thus for calcareous ooze, the rate seems to vary from 0.1 to $1 \, g/cm^2$ per thousand years in the less productive zones and from 10 to $30 \, g/cm^2$ in the more productive areas. The rate for siliceous oozes ranges from as little as $0.05 \, g/cm^2$ per thousand years in the tropics, to as high as $50 \, g/cm^2$ in the Gulf of California.

Figure 3.1 *Unconformity between Upper Cretaceous marine sediments and Archaean gneiss of the Bohemian Massif, near Kolin, Czechoslavakia (DVA)*

In shallow-water areas, of course, the rates are higher. Carbonate deposition on the Great Bahama Bank is said to have averaged half a metre per thousand years, even though most of the sediment is continuously swept into deeper water. The average rate of accumulation of the 6000 m of limestone under the Bahamas is only 4 or 5 cm per thousand years. From this disparity in rates, Professor Norman Newell deduced that these Cretaceous and Tertiary limestones represent no more than a tenth of Cretaceous and Caenozoic times.

These sediments are only slightly compactible, but with others (particularly muds) it must be realised that the thickness of new sediment is not the same thing as the thickness of the rock that results from it. Nevertheless we are always faced with a contradiction between rates of deposition and the known thickness of rock for a particular period of geological time.

When challenged with this sort of argument, most practitioners of the doctrine of continuous sedimentation then change their ground and say: 'Oh no, we just mean without significant breaks'. But what is significant? Obviously there are plenty of unconformities where the break is obvious, such as the splendid unconformity between the Upper Cretaceous and the Precambrian of the Bohemian Massif shown in Figure 3.1.

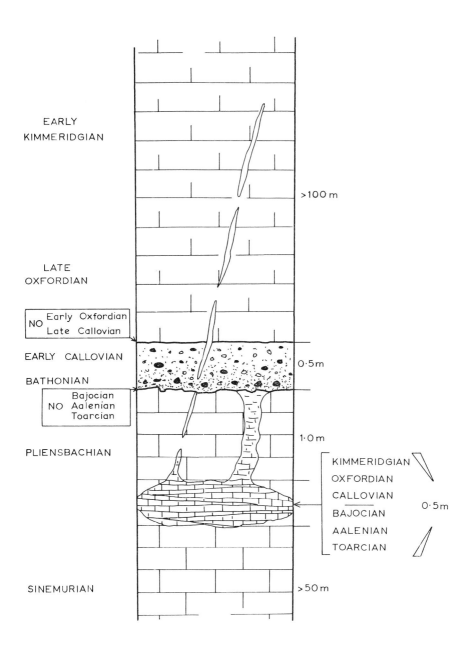

Figure 3.2 *Idealised diagram of the condensed sequences preserved in solution hollows in Jurassic limestones of west Sicily (from information in Wendt, 1965 and 1971)*

Figure 3.3   Thick, well bedded Lower Jurassic sediments above Bourg d'Oisans (Isère) in the French Alps (DVA)

However, as our studies continue, more and more concealed breaks become apparent, such as the remarkable situation in the Jurassic limestones of western Sicily, where several stages are packed away into thin solution pipes in what otherwise look like unbroken limestone sequences (Figure 3.2).

Suppose we look at some of these areas of thick 'continuous' sedimentation. Look at the spectacular cliffs of the Lower Jurassic sediment of the Dauphinois trough above Bourg d'Oisans in the French Alps (Figure 3.3). Hundreds of metres of shales and mudstones represent one small part (and I suspect one small part of that small part) of the Jurassic. Here, if anywhere, one would think we must have had continuous sedimentation. But what are all those bedding planes? What is any bedding plane if it is not a mini-unconformity? If we really had continuous sedimentation then there would surely be no bedding planes at all.

In fact the only time we see unbedded sediments, apart from comparatively small thicknesses of in-situ reef development, we can almost always find evidence of the destruction of the bedding planes by recrystallisation or by the burrowing activity of organisms. Professor Gilbert Kelling has pointed out to me that, in certain circumstances, bedding planes can be produced by textural and diagenetic differences within 'continuous sedimentation'. But I still maintain that most bedding planes show evidence of a pause in sedimentation, if not actual erosion.

Ideally, I suppose, we should look around for the thickest available development of a particular unit if we are to find anything approaching continuous sedimentation. Even then it has been calculated, taking systems as a whole, that the maximum rate of sedimentation would have been something like 300 cm in a thousand years.

In fact we have a paradox here in that the areas most commonly cited as those of continuous sedimentation without breaks, such as the late Ordovician—early Silurian of the Southern Uplands of Scotland and the Jurassic to early Oligocene of the Italian Apennines, are also those of thinnest sedimentation. Clearly, in such developments there may be few, if any, erosional breaks, but there must be immense non-depositional breaks. And even in deposits such as the flysch of northern Spain or the Polish Carpathians, there is a great deal of evidence of erosion by the turbidity currents that laid down the sediment.

Having a sentimental attachment for them, I cannot resist mentioning the Cotswold Hills of western England where the formation still quaintly known by William Smith's original name of the 'Inferior Oolite', reaches what is for us the tremendous thickness of about 30 m. These are the Aalenian and Bajocian Stages of the Middle Jurassic and compare favourably with the equivalent on the Dorset coast of southern England, where the limestones of this age are condensed into a mere 3.4 m with two obvious breaks (Figure 3.4). But it has long been known from careful palaeontological studies that even in the thick development in the Cotswolds, there is evidence of two major breaks and a period of folding in what otherwise was a very peaceful period in British geological history. What is more, if we look farther afield, our magnificent 30 m dwindles into insignificance. In Alaska we read that the Middle Bajocian alone amounts to more than 1200 m.

Again, the child-like wonder appears when we read for example of more than 2000 m of Kimmeridgian (Upper Jurassic) in New Zealand or more than 3000 m of Frasnian (Lower part of the Upper Devonian) in Arctic Canada, or more than 5000 m of Arenigian (Lower Ordovician) in western Ireland. What I think of as a few steps along the beach in the Isle of Wight suffices

Figure 3.4   *Condensed deposit of Middle Jurassic limestones, Burton Bradstock, Dorset, England, showing breaks (DVA)*

to pass the Middle Oligocene, but I find this amounts to untold thousands of metres of sediment in New Guinea.

For any tiny part of the stratigraphical column of which we are particularly fond in our own backyard, we can almost always find somewhere else in the world where the same division is a hundred or a thousand times thicker. We are only kidding ourselves if we think that we have anything like a complete succession for any part of the stratigraphical column in any one place.

Even within a short distance we can see remarkable changes of thickness. The so-called 'Purbeckian' Stage at the top of the Jurassic reaches a maximum of more than 170 m under the Weald, south of London. Yet only 48 km or so away the 'Purbeckian' is down to less than a metre with no obvious breaks, on the French cliffs near Boulogne. Very often it is not as simple as that. What looks like thinning in a major unit may turn out to be something much more complicated in the smaller constituent units.

The textbook picture of the lower part of the Lower Jurassic in England, looked at in the usual two-dimensional textbook way, along the outcrop, is of thinning over three axes with thicker

49

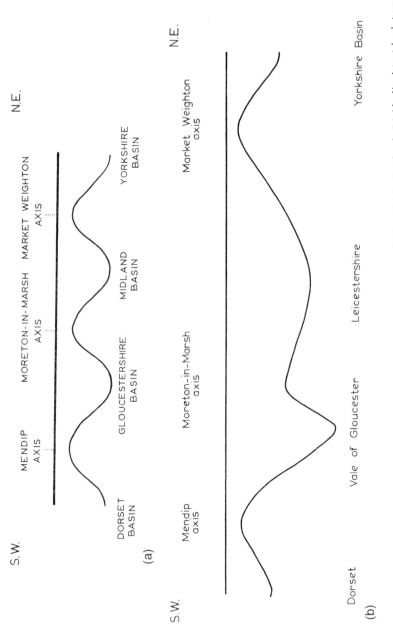

Figure 3.5 Contrasted versions of the variation in thickness of the lowermost Jurassic deposits in England: (a) idealised text-book type variation along the outcrop showing 'axes' and intervening 'basins'; (b) the same variation expressed in actual total thicknesses

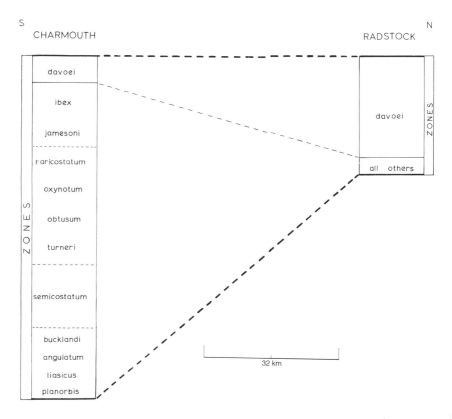

Figure 3.6 *Actual variation within constituent zones between the Dorset coast and the Mendip 'axis'*

basins of sedimentation in between (Figure 3.5a). If we put in actual thicknesses, the 'axes' are not so obvious, but they are still there (Figure 3.5b). However, if we look at just one part of that story in detail, we find complications. Thus if we trace the zones of the Lower Lias (as it is called) from the Dorset 'basin' to the Mendip 'axis', we find that 11 of the 12 zones thin as they ought (Figure 3.6) with no signs of any breaks in the succession, but the twelfth actually thickens markedly!

Such misbehaviour of strata in their most classic sections leads me to have serious doubts (in fact, positive hatred) of the concept of the 'stratotype' so much favoured by many workers on the European continent. This idea of type section for a particular stratigraphical division will be discussed in a later chapter; all I must say here is that no type section known to me can possibly

Figure 3.7 *Stratigraphical break with bored phosphatic nodules in the 'stratotype'
of the Volgian (uppermost Jurassic) at Gorodishchi near Ulyanovsk, USSR (DVA)*

pretend to be representative of a whole unit of the stratigraphical
column, however small. Keeping, naturally enough, to the
Jurassic (since it was the birthplace of stratigraphy), let us take
as an example the Volgian Stage. Previously this competed with
the Tithonian (discussed in Chapter 1) for a place at the top of
the Jurassic. A lesser rival was the Portlandian (cum 'Purbeckian')
of England, to say nothing of the quaintly named Bononian and
Bolonian of France.

The 'stratotype' of the Volgian Stage is at Gorodishchi, along
the river from the birthplace of Lenin (Ulyanovsk, formerly
Simbirsk). It is an excellent section, packed with fossils, but with
very obvious breaks marked by prominent bands of phosphatic
nodules and borings (Figure 3.7). Clearly any break is a
disadvantage in a section that sets out to typify a whole division

of geological time, but here it is worse, for one of those breaks is now thought (at least by one eminent British palaeontologist) to conceal the whole of the British Portlandian.

All this leads me to the conclusion that the greater part of the passage of geological time has left over most of the earth no more than Shakespeare's 'gap in nature'. If you study textbooks or correlation charts, such as those produced by the Geological Society of America and the Geological Society of London, you come inevitably to the conclusion that the stratigraphical column in any one place is a long record of sedimentation with occasional gaps. Even though we may pay lip service to the idea that there are probably many breaks not yet discovered, nevertheless almost every geologist seems to accept the above doctrine, albeit subconsciously.

But I maintain that a far more accurate picture of the stratigraphical record is of one long gap with only very occasional

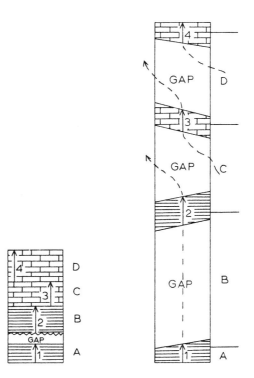

Figure 3.8   Comparison of the conventional picture of a particular part of the strati-graphical record (left) with what is probably the true picture (right); see text for explanation

sedimentation. This doctrine is illustrated in Figure 3.8. On the left you see the conventional picture of part of the stratigraphical columns, with varied sediments, a single small break and fossil records (indicated by numbers) that start from nothing and end as nothing. On the right you see what I think is a far more likely explanation of the same facts. The gaps predominate (and probably should be far longer than they are shown here), the lithologies are all diachronous and the fossils migrate into the area from elsewhere and then migrate out again.

Perhaps the best way to convey this attitude is to remember a child's definition of a net as a lot of holes tied together with string. The stratigraphical record is a lot of holes tied together with sediment. It is as though one has a newspaper delivered only for the football results on Saturdays and assumes that nothing at all happened on the other days of the week. To change my metaphor yet again, I would compare the stratigraphical record with music. Just as the intervals between the notes in music are every bit as important as the notes themselves, so the bedding planes are as important as the beds.

No doubt my prejudices are coloured by having looked at too much epicontinental sediment and not enough oceanic, but I must plead in my defence that this is the nature of the stratigraphical record on the continents anyway. What is more, we now know from the Deep Sea Drilling Project that there are vast gaps in the record of the oceans, too.

So we may come to the third proposition of this book: *the sedimentary pile at any one place on the earth's surface is nothing more than a tiny and fragmentary record of vast periods of earth history.* This may be called the 'Phenomenon of the Gap Being More Important than the Record'.

REFERENCES

Casey, R. (1968). 'The Type-Section of the Volgian Stage (Upper Jurassic) at Gorodische, Near Ulyanovsk, USSR', *Proc. Geol. Soc. Lond.*, No. 1648, pp. 74–75.
This brief note, together with one the previous year in the same publication, illustrates the inadequacy of the stratotype concept.
Lisitzin, A. P. (1971). 'Distribution of Siliceous Microfossils in

Suspension and in Bottom Sediments' *and* 'Distribution of Carbonate Microfossils in Suspension and in Bottom Sediments', in B. M. Funnell and W. R. Riedel (eds), *The Micropalaeontology of Oceans.* Cambridge University Press, Cambridge, pp. 173–195 and 197–218.
A survey of the rates of accumulation of oceanic organic ooze.

Newell, N. D. (1967). 'Paraconformities', in *Essays in Paleontology and Stratigraphy. R. C. Moore Commemoration Volume. Univ. Kansas Dept. Geol. Spec. Publ.*, No. 2, pp. 349–367.
Emphasises the importance of sedimentary breaks of regional extent, which are not obvious in the field, but represent significant time intervals.

Ramsay, A. T. S. (1977). 'Sedimentological clues to palaeo-oceanography', in A. T. S. Ramsay (ed.), *Oceanic Micropalaeontology,* Vol. 2, Academic Press, London, New York, pp. 1371–1453.
A clear demonstration of the great gaps that occur even in the sediments of the deep oceans.

Sujkowski, Z. (1931). 'Petrografia kredy Polski. Kreda z głebokiego wiercenia w Lublinie'. *Spraw. Panstw. Inst. Geol., Warszawa,* Vol. 6, pp. 484–628.
The original description of the very thick Upper Cretaceous of the Lublin bore-hole in south-east Poland.

Wendt, J. (1965). 'Synsedimentäre Bruchtektonik im Jura Westsiziliens', *N. Jb. Geol. Paläont. Mh.,* Vol. 5, pp. 286–311.

Wendt, J. (1971). 'Genese und Fauna Submariner Sedimentärer Spaltenfüllungen im Mediterranen Jura', *Palaeontographica Abt. A,* Vol. 136, pp. 122–192.
Two accounts of the remarkable condensed deposits in solution crevices within Jurassic limestones in west Sicily.

4
Catastrophic Stratigraphy

I have heard it said that the only major advance in stratigraphy and sedimentology since the Second World War was the concept of the turbidity current. This is probably true, but it seems to me that we have now become so engrossed in the jargon of 'turbidites' (an objectionable genetic term) and with their 'bottom structures', that most of us have lost sight of the real significance of the original idea.

In effect, in one step the concept of the turbidity current took us back to the catastrophism of earlier geological thought. In place of the comfortable doctrine of sedimentation keeping pace with subsidence, we were now faced with the notion of empty troughs forming in the sea-floor, with only occasional rushes of sediment to fill them.

Thus in the flysch of the Polish Carpathians (Figure 4.1), it has been estimated that there was, on average, one turbidity current flow every 29 000 years. In the even more classic ground of the Italian Apennines, where Kuenen and Migliorini first recognised the turbidity current, single graded beds up to 20 m thick within the Macigno Sandstone are said to have been deposited by a single 'whoosh' of turbid water.

What is more, the Macigno Sandstone as a whole represents a startling change in sedimentation from what went before. In the Apennines, the whole of the Upper Jurassic, the whole of the Cretaceous, the whole of the Paleocene and Eocene and the lower part of the Oligocene—in an apparently unbroken succession—may amount to as little as 70 m of sediment. But then

Figure 4.1   *Lower Cretaceous flysch near Bielsko Biala in the Polish Carpathians (DVA)*

the rest of the Oligocene alone reaches as much as 3000 m. This is the Macigno Sandstone. The story is that slow sedimentation in deep water produced first the *Diaspri* (Upper Jurassic radiolarian cherts) and then the *Scaglia* or *Scisti policromi* (Cretaceous to Oligocene vari-coloured shales). Then, at the top of the *Scaglia*, there was a sudden change in sedimentation to a brecciated limestone full of large foraminifera (the *Brecciole nummulitiche*). This heralded the incomparably faster deposition of the Macigno 'turbidites'.

Even more 'catastrophic' deposits were to follow in the Apennines, for with the earth movements indicated by the synorogenic sediments mentioned above, there came into Italy from the west an allochthonous series of nappes. This was formerly known as the *Argille scagliose* or 'scaly clays'. It was thought to be nothing more than a shattered argillaceous deposit, produced by repeated submarine landslipping and carrying with it exotic blocks ranging in size from small fragments (Figure 4.2) up to whole mountains. Some of these are unknown in the autochthonous succession of the Italian mainland, for example the great masses of serpentinite and associated rocks seen near the west coast. Their size diminishes as they get further away

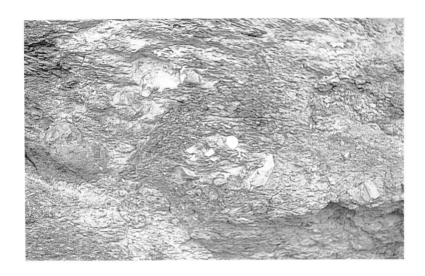

Figure 4.2 *Exotic blocks in the* argille scagliose *Passo sec, Bracco in the Italian Apennines*

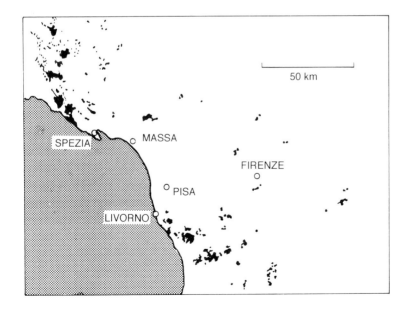

Figure 4.3 *Distribution of allochthonous ophiolites in the north Apennines (after Merla, 1952)*

from their source area in the Tyrrhenian Sea (Figure 4.3). More recent work has demonstrated that the allochthon is not quite as confused as was formerly thought, and the term *Argille scagliose* has been dropped in Italy (though it is still used in many other parts of the world). The poetic name has been replaced by terms such as 'chaotic complex' and the pseudo-scientific 'olistostromes' without really adding to our knowledge, but the essential point is that submarine landslipping is now thought to be only part of the process rather than the whole. Modern thought has reverted to earlier nappe theories in which sizeable slices of recognisable stratigraphy have been pushed forward considerable distances from the south-west. At the same time there has been extensive sliding, especially at the front of the nappes. But rather than reducing the amount of such sliding, modern ideas seem to have increased it, for olistostromes are now recognised in all parts of the succession, autochthonous as well as allochthonous.

This example was worth discussing in more detail than most because the *Argille scagliose* type of 'catastrophic' deposit is now recognised in many parts of the world, from California to Taiwan. In Turkey, near the Chalk cliff mentioned at the beginning of Chapter 1, there is a great jumble of Cretaceous blocks resting, with a very irregular junction, on Eocene nummulitic marls. This had been described as a thrust, but it is almost certainly a submarine landslip deposit like those in Italy. In fact, many years ago I was rash enough to suggest that the great Ankara *mélange* itself (humorously called by the locals *türlü güvec*—a Turkish version of Lancashire hotpot) might be an *Argille scagliose* type of deposit. I was duly slapped down by my more knowledgeable tectonic seniors, and tried to forget the brief publication in question, but later work has now led met to suspect that I might have been right after all.

Certainly there are many smashed-up-looking deposits around the world carrying huge exotic blocks far from their place of origin. One of the most remarkable I have heard of is in the island of Timor, where there is a deposit, known as Bobonaro Scaly Clay, which extends for nearly 1000 km of outcrop, is nearly 100 km wide and is 2.5 km thick. In fact, if its alleged occurrence on other islands is correct, it extends for at least 1600 km. Figure 4.4 shows a rounded exotic pebble in this deposit, with some

Figure 4.4   *Exotic rounded block in the Bobonaro Scaly Clay, near Bobonaro in eastern Timor; note the men providing a scale at the bottom of the block (photograph kindly provided by Professor M. G. Audley-Charles)*

men at the bottom to provide a scale. Quite apart from the problem as to how this massive chunk of basic igneous rock became rounded, it is difficult to use a term other than 'catastrophic' for the arrival of such a pebble. It is interesting to note, in passing, that rather similar chaotic deposits and slump topography have now been found at the foot of many present-day continental slopes.

However, it is not necessary to go to the remoter parts of the world to find examples of such deposits. In stratigraphy one is always using one ruler to measure another and one can only call a deposit exceptional if we have something more 'normal' with which one can make a comparison.

Thus, though the placid British Jurassic sediments have received vastly more than their fair share of study, it is curious that some of their more spectacular features have been relatively neglected. The minutiae of its English outcrop from Dorset to Yorkshire have probably been turned over by more loving hands than have touched any equivalent heap of sediment elsewhere. The narrow outcrops on either side of Scotland have been honoured by comparatively few publications, though their story is much more exciting. Thus, while we follow with interest the supposedly dramatic variation from 60 cm to 3 m to 12 m of a certain zone along the English outcrop, we hardly bother with a footnote to say that the same zone in western Scotland is more than 40 m thick.

On the east side of Scotland, in Sutherland, there are what are—to me—the most exciting deposits in the British Jurassic. The Kimmeridgian stage of the Upper Jurassic is here developed to an exceptional thickness in a narrow strip along more than 16 km of coast. Interbedded in the usual deepish-water black shale are numerous boulder beds, that have been variously interpreted as ordinary sedimentary conglomerates, as tectonic breccias, as deposits from melting ice and as rock falls from towering cliffs. Sir Edward Bailey's lucid explanation of these deposits was that the boulders (up to more than 30 m long) fell from a submarine fault scarp, probably triggered by earthquakes which also produced the clastic dykes that are a feature of the sections. Each seismic shock produced a tsunami, which swept shallow-water sediments and fauna (including reef corals) down the fault scarp to settle among the boulders. Bailey's interpretation of this 'natural seismograph', as he called it, is certainly applicable in many other places around the world.

The gigantic Tertiary boulder beds of Ecuador are perhaps the best-known example. Boulders up to 3 km long are said to have fallen down a scarp along an outcrop more than 300 km long. The Tertiary 'Wildflysch' of Switzerland may be another example. I have used this kind of explanation myself for repeated boulder beds within a thick Cretaceous limestone sequence near Lagueruela in Teruel Province, in eastern Spain.

All these examples provide clear evidence of very rapid sedimentation of a 'catastrophic' kind in what may be regarded as exceptional deposits in areas of tectonic instability. But let us also remember that Baron Cuvier, perhaps the greatest protagonist of catastrophism in its true form, worked on the placid, untectonised sediments of Montmartre and the environs of Paris. The phenomenon I am trying to demonstrate in this chapter is seen in the most domestic of deposits.

Thus, the Neogene sediments around the Mediterranean are, for the most part, flat-lying shallow-marine and richly fossiliferous. Yet recently Dr Hsü of Zurich has suggested that they record what must have been one of the most spectacular incidents in earth history. At the end of the Miocene, the Mediterranean became a deep, desiccated basin with evaporites and lacustrine diatomites. At the beginning of the Pliocene, the Strait of Gibraltar opened again, according to Hsü, as a fantastic cataract through which a catastrophic deluge refilled the Mediterranean. Hsü (1983) estimated that this was more than a thousand times greater than Niagara, but still took more than 100 years to refill the Mediterranean.

In the next chapter I shall discuss the effects of present-day violent phenomena such as hurricanes. In the course of work on the Mesozoic rocks of the High Atlas mountains of Morocco, I came across some deposits in the Upper Jurassic near Imouzzer-des-Ida-ou-Tanane which I could only interpret as having resulted from the effect of storms on lagoonal sediments (Figures 4.5 and 4.6). Bed after regular bed shows a dark, laminated lower portion which I identify as an algal mat type deposit that has been ripped up and incorporated, in a graded fashion, in 'cleaner', paler sediment from offshore. Quite independently, my former colleague Gilbert Kelling had come to the conclusion

Figure 4.5 *Supposed storm deposits in Upper Jurassic near Imouzzer-des-Ida-ou-Tanane, western High Atlas, Morocco; each of the beds represents a distinct storm episode, with black carbonate below and paler carbonate above (DVA)*

Figure 4.6 *Close-up of one of the above units; black laminated deposits (below) algal mats, possibly ripped up and graded in the 'cleaner' sediment from off-shore (above); the coin is a dirham, 2.4 cm across (DVA)*

that certain sediments in the stratigraphical record can be best interpreted in terms of violent storms. By analogy with 'turbidites' he has coined the term 'tempestites' for graded beds of shallow water sediments that may have been churned up by storms and allowed to settle again. They differ from turbidites in their general environmental setting and in the relative paucity of the basal sole marks that record the erosional passage of a turbid current. Professor Kelling used the term first for Carboniferous deposits on the Moroccan Meseta, but I have seen the same sort of thing in Jurassic carbonates on the Polish foreland and they have been described from many other horizons and areas, though not previously interpreted in this way. In fact the literature is now full of storm deposits and my students are finding them everywhere (without any pressure from me). One

Figure 4.7 *Supposed 'tempestite' in Lower Carboniferous limestone near Canfranc Estacion in the Spanish Pyrenees; angular black fragments are seen in a pale grey matrix, the white calcite cracking is restricted to the derived blocks, which probably represent older shallow-water algal deposits; the hammer is 29 cm long (DVA)*

Catskill delta, a flash-flood (itself an example of a modern catastrophic event) uncovered a whole forest of in-situ Devonian trees up to 12 m high.

By such means it is possible, within the Lancashire Coal Measures for example, to demonstrate that very rapid sedimentation alternated with very slow sedimentation and that the former was responsible for the bulk of at least some parts of the record.

If we turn to volcanic deposits, which can hardly be regarded as exceptional in earth history, we can find many examples of great thicknesses accumulating very rapidly indeed. At Builth in Central Wales, a complicated history has been worked out for one part of the Ordovician. First, spilitic lavas were extruded, layer upon layer, and weathered to produce a staircase-like scenery (or 'trap topography'). Their weathering gave rise to sandy and pebbly deposits. Then a series of keratophyres were extruded on top of the spilites and weathered into rounded hills with detrital pyritous sand banked up against them. Finally the sea encroached on this topography producing steep cliffs, inlets, sea stacks and sandy or shingly beach deposits. A reconstruction of the supposed scenery of this time is given in Figure 4.8 (it will be noted that the view is looking westwards and that the shadows are therefore coming from the north, evidently proving that the British Isles were then in the southern hemisphere!). But the most surprising fact about this is that all these events took place during the deposition of a single graptolite zone. Admittedly Ordovician graptolite zones must have lasted much longer than the ammonite zones of the Mesozoic, but nevertheless the time scale is still surprising.

The great problem of uniformitarianism was always the amount of time needed to explain what was known to have happened in the history of the earth (including the evolution of all its species) if one could only postulate present processes. James Hutton recognised this right from the start with his famous aphorism 'no vestige of a beginning, no prospect of an end'. Charles Lyell was a student of William Buckland who taught that 'Geology is the efficient auxiliary and handmaid of religion' and who saw evidence everywhere of 'direct intervention by a divine creator', of a 'creative power transcending the operation of known laws of nature'. Lyell instead wanted an empirical theory

Figure 4.8 *Reconstruction of the supposed scenery in early Ordovician times, near Builth, Wales (from Jones and Pugh, 1949, by kind permission of Sir William Pugh and the Geological Society of London)*

founded on observation and practical experience that did not depend on any preconceived ideas from the Bible or elsewhere and which provided a methodology—a means of explaining geological phenomena.

'Methodological uniformitarianism', as it is sometimes called, makes the simple assumption (as in all other sciences) of the invariance of natural laws. 'Substantive uniformitarianism' on the other hand (towards which Lyell also inclined) presumes uniform rates or conditions. It is this second concept which has caused all the trouble, certainly (for example) in view of what we now know about the very different world of early Precambrian times. But what chiefly concerns me here is simply the amount of time needed for what we know to have happened in the geological past if we can only postulate 'normal processes'. Thus the other great uniformitarian of the last century—Charles Darwin—estimated vastly longer periods than are accepted today. For example, he postulated 300 million years since the Cretaceous to allow for the scooping out of the Wealden anticline in south-east England.

Uniformitarianism triumphed because it provided a general theory that was at once logical and seemingly 'scientific'.

Catastrophism became a joke and no geologist would dare postulate anything that might be termed a 'catastrophe' for fear of being laughed at or (in recent years) linked with a lunatic fringe of Velikovsky and Californian fundamentalists. But I would like to suggest that, in the first half of the last century the 'catastrophists' were better geologists than the 'uniformitarians'. Baron Cuvier in France was a remarkable character and made very detailed studies of the sediments and fossils of the Paris Basin that were years ahead of his time. He recorded what was really there in the rocks, that is to say repeated and sudden changes in environments and extinctions of animals and plants. He was not a crank, but an extremely careful observer. (It is generally forgotten that whereas Cuvier was really the first to recognise the extinction of past species, Lyell clung to the notion that the earth, together with its complete flora and fauna, had always been essentially as it is now.) Cuvier's counterpart in Britain was a much more prosaic son of the Industrial Revolution—William Smith—whose conclusions were basically very similar but who was far too much a practical man to want to theorise about it.

The middle years of the last century were dominated in British geology by Sedgwick and Murchison, who both supported catastrophic doctrines all their lives (though not in a strictly biblical sense as had Buckland before them). In many ways these two symbolise the 'old order' in the British social system—Adam Sedgwick the high churchman (like Buckland) and Roderick Murchison the Tory landowner and ex-soldier from Waterloo.

On Charles Lyell's side at first there was no one of similar stature, though one might mention the great Goethe in Germany who was a keen amateur geologist and liked the gradual peaceful processes preached by the uniformitarians.

The long battle that followed matched the political and philosophical situation of the time. On the catastrophists' side were the Tories and the Church (the episcopalian Church of England, of course). They supported the idea of monarchy as the natural state of things, with the monarch ruled only by God, the divine monarch, who controlled the day-to-day happenings on earth, geological as well as human. The history of that earth was recorded in the Bible. Though it is so far from the world in which most of us live today, it is difficult to over-emphasise the strength of the Church and the landed gentry at that time. Captain Fitzroy of the *Beagle*, who argued so much with Darwin, was a nephew of Lord Castlereagh, that haughty aristocratic Tory minister of Napoleonic days, and himself became a Tory MP. All his life he was a staunch supporter of the Bible as the ultimate truth and proclaimed this at the famous Oxford debate on evolution.

On the uniformitarian side, in my glorious over-simplification, were the Whigs or Liberals who sought to dispute the absolute rule of a monarchy as the natural state of affairs and were in favour of majority rule. They proclaimed the liberal sciences as demonstrating that Nature was governed by natural, not supernatural, processes; they were in favour of gradual change.

Like good historians we should try to look at their world through their eyes. 'Liberals' in the broad sense, including most of the scientists, thinkers and poets of the day, tended to be linked in people's minds with the excesses of the French Revolution and the nearly 30 years of war that followed it. (We must remember that this was closer to them than the Second World War is to us.) Besides the politics, of course, the liberal scientists were also linked with anti-religion. That great Scottish pioneer Hugh Miller opens his book on the Old Red Sandstone with advice to young men not to attend Chartist meetings, but to read the Bible and study geology. It is a curious paradox that evolution and gradual change were linked with revolution and sudden change.

Charles Lyell in fact had to foreswear his beliefs like Galileo to get the chair of geology at that holiest of London Colleges— King's. Later, ladies were forbidden to attend his lectures because of his shocking views and he resigned (though it is not recorded if he missed the ladies or just their fees).

It is not, I think, a coincidence that the publication of Lyell's *Principles of Geology* coincided with the Great Reform Act of 1832. In the same year Darwin set sail on the *Beagle*, taking with him Lyell's *Principles* and reading it so much that he had to ask the ship's carpenter to rebind it in wood. (Much later in the century Lenin had a copy of it, too, in his study bedroom in Simbirsk.) Of course, evolution by natural selection was completely uniformitarian in attitude. Darwin deduced that it happened in the past from what he saw happening today—as in the finches and tortoises of the Galapagos Islands. Hutton had in fact summarised the principles of natural selection in unpublished manuscript notes way back in the 18th century.

Eventually the uniformitarian cause won—or so it seemed— and catastrophism became an old-fashioned joke. In fact I must emphasise again that, in my view, the early uniformitarians were the theoreticians and the catastrophists were the careful field observers. Lyell himself was an excellent man on principles and processes, but he did not take very much interest in rocks. There is an interesting letter from Scrope to Lyell written in 1832 (after the publication of the second volume of the *Principles*). In it he says:

> It is a great treat to have taught our section-hunting quarry men that two thick volumes may be written on geology without once using the word 'stratum'.

My excuse for this lengthy and amateur digression into history is that I have been trying to show how I think geology got into the hands of the theoreticians who were conditioned by the social and political history of their day more than by observations in the field. So it was—as Steve Gould put it—that Charles Lyell 'managed to convince future generations of geologists that their science had begun with him'.

In other words, we have allowed ourselves to be brain-washed into avoiding any interpretation of the past that involves extreme and what might be termed 'catastrophic' processes. However, it seems to me that the stratigraphical record is full of examples of processes that are far from 'normal' in the usual sense of the word. In particular we must conclude that *sedimentation in the past has often been very rapid indeed and very spasmodic*. This may be called the 'Phenomenon of the Catastrophic Nature of much of the Stratigraphical Record'.

REFERENCES

Abbate, E., Bortolotti, V., Passerini, P. and Sagri, M. (1970). Introduction to the Geology of the Northern Apennines', *Sediment. Geol.,* Vol. 4, pp. 207–249.
This, and the whole series of papers that immediately follow it, give

an admirable summary in English of modern views on the northern Apennines including the 'olistostromes' and similar deposits.

Ager, D. V. (1956). 'Summer Field Meeting in Italy', *Proc. Geol. Ass.*, Vol. 66, pp. 329–352.
A general account in English of the remarkable geological history of the Apennines, as formerly interpreted by Migliorini and Merla.

Ager, D. V. (1974). 'Storm Deposits in the Jurassic of the Moroccan High Atlas', *Palaeogeog., Palaeoclimat., Palaeoecol.*, Vol. 15, pp. 83–93.
One of dozens of papers I could cite which give evidence of storm-deposits in the stratigraphical record, though one by which I am particularly convinced. It lacks the conviction of photographs (due to a strike) so these are provided herein (Figures 4.5 and 4.6).

Audley-Charles, M. G. (1965). 'A Miocene Gravity Slide Deposit from Eastern Timor', *Geol. Mag.*, Vol. 102, pp. 267–276.
An account of an *Argille scagliose* type deposit with the boulder shown in plate IX (see Figure 4.4 this volume).

Bailey, E. B. and Weir, J. (1932). 'Submarine Faulting in Kimmeridgian Times: East Sutherland', *Trans. Roy. Soc. Edinb.*, Vol. 57, pp. 429–467.
The classic paper on the Sutherland boulder beds.

Birkenmajer, K. and Gasiorowski, S. M. (1961). 'Sedimentary Character of Radiolarites in the Pieniny Klippen Belt, Carpathians', *Bull. Acad. Polon. Sci.*, Vol. 9, pp. 171–176.
For interesting calculations on the frequency of turbidity currents.

Broadhurst, F. M. and Loring, D. H. (1970). 'Rates of Sedimentation in the Upper Carboniferous of Britain', *Lethaia*, Vol. 3, pp. 1–9.
A follow-up from earlier papers in which the authors demonstrate the alternation of very slow with very rapid sedimentation during Late Carboniferous times.

Conybeare, C. E. B. (1967). 'Influence of Compaction on Stratigraphic Analysis', *Bull. Canad. Petrol. Geol.*, Vol. 15, pp. 331–345.
An essential reference in all these discussions, dealing with the vital question of the relative compactability of sediments.

Hawkins, H. L. (1953). 'A Pinnacle of Chalk Penetrating the Eocene on the Floor of a Buried River-Channel at Ashford Hill, Near Newbury, Berkshire', *Quart. J. Geol. Soc. Lond.*, Vol. 108, pp. 233–260.
An account of the surprising things that can happen in the most placid of terrains.

Hsü, K. J. (1983). *The Mediterranean was a Desert. A Voyage of the* Glomar Challenger. Princeton University Press, Princeton, NJ.
A very readable blow-by-blow account of this remarkable discovery.

Jones, O. T. and Pugh, W. J. (1949). 'An Early Ordovician Shore-line in Radnorshire, Near Builth Wells', *Quart. J. Geol. Soc. Lond.*, Vol. 105, pp. 65–99.

Two very eminent British geologists combine in a fascinating account of a remarkable area.

Kerney, M. P. (1963). 'Late-Glacial Deposits on the Chalk of South-East England', *Proc. Roy. Soc. (B)*, Vol. 246, pp. 203–254.
Demonstrates quite sudden environmental changes in geologically recent times.

Kuenen, P. H. and Migliorini, C. I. (1950). 'Turbidity Currents as a Cause of Graded Bedding', *J. Geol.*, Vol. 58, pp. 91–127.
The classic paper on turbidity currents as a major factor in the formation of geosynclinal sediments.

Kulp, J. L. (1961). 'Geological Time Scale', *Science, N.Y.*, Vol. 133, No. 3459, pp. 1105–1114.
A useful survey of the radiometric dating, including a detailed subdivision of the Cretaceous that has since been disputed.

Merla, G. (1952). 'Geologia dell'Appennino Settentrionale', *Boll. Soc. Geol. Ital.*, Vol. 70, pp. 952–982.
The standard account of the geological history of the Apennines.

5
Catastrophic Uniformitarianism

It follows naturally from the previous chapter that we should now go on to consider where sedimentation is actually taking place today. Here it is important to distinguish between ephemeral sedimentation that comes and goes with the seasons and permanent sedimentation that actually accumulates and stays. My main complaint concerning students of modern sediments is that they pay more attention to the question of how sediments are deposited than to the question as to whether or not they stay there. Thus the excellent studies of the mud and sand flats around the Wash in eastern England give us a clear picture of the zonation of the sediments and their associated life, and of the processes involved, but do not tell us if they have accumulated in depth and are likely to stay there for the geologist of the future. In fact there is clear evidence that the meandering creeks that wander across the flats are eating away the sediment again soon after it is deposited (Figure 5.1). It is also well known that the sediments build up in the summer and tend to be removed again in the winter. Thus, though there is a slow progressive build-up in the upper levels of the intertidal zone, largely aided by human interference, in the lower levels there is virtually no overall accretion.

If we look at the sea-floor maps that are now becoming increasingly available, one is struck (at least, I am) by the great areas that are either receiving no sediment at all or else are covered with the merest veneer. Thus it has been written

Figure 5.1   *Laminated sediment being eroded at the side of a creek in the Wash, eastern England (DVA)*

recently: 'The shelf off eastern United States is covered almost entirely with relict nearshore sands of the (Pleistocene) transgression.'

Even where sediment is recorded, it is frequently in the form of sand waves that move from place to place and do not accumulate. Most of the sediment in fact seems to be accumulating close inshore and very little gets to the outer shelf or the deeps. It has been calculated that there has been an average of about 9 m of deposition close inshore during the last 5000 years. Coupled with this, however, we have to remember the huge concentration of such sediment in deltas such as that of the Mississippi, where it has been accumulating at a fantastic rate (perhaps 3000 m in the same period of time). Similarly, around the mouth of the Orinoco in Venezuela, the area of rapid sediment accumulation is remarkably limited. Beyond the shelf, accumulation seems to be concentrated in a comparatively few deep-water basins. All this, of course, is during a time of exaggerated relief following the Pleistocene glaciation. It is for these reasons that sedimentologists have been forced to work to death the few modern examples they have (such as the poor old Bahamas Banks) for analogies with ancient sedimentation.

Years ago, Arthur Holmes made an interesting calculation dividing the present area of sea-floor by the total amount of sediment being brought down annually by all the rivers of the world. He estimated 8×10^9 tonnes transported annually to the sea, which works out at 0.025 kg per square metre of the sea-floor. If the average density of the sediment is 2000 kg/m^3, the average rate works out at about 1 cm per thousand years. This is even less than some of the rates mentioned in Chapter 3.

It seems to me, from a number of recent papers (and from common sense) that the rare event is becoming more and more recognised as an important agent of recent sedimentation. Papers have been written on 'the significance of the rare event in geology' and one must never forget the significance of the old truism that given time, the rare event becomes a probability and given enough time, it becomes a certainty. We certainly have enough time in geology. A study of the 1961 hurricane 'Carla' and the 1963 hurricane 'Cindy' in the southern United States showed that they had considerably modified both the form of the affected coastline and the distribution of sediments there. The suggestion was that just as the energy of electrons is discharged in discrete amounts, or *quanta*, so energy is expended in near-shore sedimentary environments within short time intervals that are separated by long periods of relative calm. In other words, the changes do not take place gradually but as sporadic bursts, as a series of minor catastrophes.

It has been calculated that, in the Gulf of Mexico, there is a 95% probability that a hurricane will pass over a particular point on the coast at least once in 3000 years. The maximum amount of sediment likely to be deposited over that period along the coast generally is about 30 cm and we know that hurricanes will certainly rearrange that amount of material. In other words, the rare hurricane is likely to be the main agent recorded in the stratigraphical column of certain parts of the world, even in our present climatic set-up.

Similarly, it has been shown that tsunamis, or 'tidal waves' as they were for a long time mis-named, have an immense effect on shorelines, both in erosion and in the shifting of great quantities of sediment. To quote a recent author on the subject: '. . . the action of tsunamis is short and extremely violent . . .'. It has been suggested that sea-floor sediments as deep as

1000 m may be disturbed. Waves up to 40 m high have been recorded rushing inland, carving out valleys, stripping off deltas and wiping out hills. The resultant mass of land, beach and shallow-water sediments is just as violently carried out to sea and dumped.

In the Aleutian island of Unimak on the 1 April 1946, a tsunami produced by a local submarine earthquake swept away not only a massive lighthouse on a promontory nearly 10 m above the sea, but also a radio mast and coastguard station with 20 men more than 30 m up. The same waves reached Hawaii in the central Pacific less than 5 hours later and must have travelled at a speed of 740 km/h.

It is commonly said that tsunamis are usually triggered by earthquakes or violent volcanic explosions. It is also possible that they can be produced by the slumping of large masses of sediment in water, though in this case the cause may be confused with the effect.

Though infrequent, there are certainly enough of them for geological purposes. From historical records it can be deduced that there have been more than 200 notable tsunamis in the last two thousand years; this would allow us more than 100 000 in a million years. It is also noteworthy that they are most effective on the steeply sloping shores of tectonically active regions, such as Japan, where they got their name. Though their amplitude is low, they tend to be damped into ineffectiveness on wide continental shelves.

This association, both through cause and effectiveness, with tectonically active regions may be significant in view of their alleged association with turbidity currents and chaotic deposits. They may, in fact, explain many of the curious features of so-called 'turbidite' and similar sequences. Probably the first time that tsunamis were blamed for a particular deposit was in Sir Edward Bailey's brilliant exposition of the origin of the Upper Jurassic boulder beds in eastern Scotland, discussed in the last chapter.

One could go on almost indefinitely finding examples of sudden dramatic natural events within the memory of man that help to explain some aspects of the geological past. The disastrous Lisbon earthquake of 1755 not only shook that city and the faith of the 'Age of Reason' (including Voltaire's ever hopeful Candide), it also considerably modified the local sea-floor and its sediments.

Perhaps the most remarkable example of all is that of the floods from the glacially-dammed Lake Missoula in Montana, described by that grand old man of American geology, J. H. Bretz. Although this has ben argued over for 50 years, the size of this ancient catastrophe now seems incontrovertible. What is more, it is close enough to the present day to be regarded as an illustration of the present-day processes by means of which we interpret the past.

Bretz's last paper on the subject sums it up in vivid terms:

> Although paleo-Indians probably were already in North America, no human ear heard the crashing tumult when the Lake Missoula glacial dam . . . burst and the nearly 2000 foot [610 m] head of impounded water was free to escape from the Clark Fork River valley system of western Montana and across northern Idaho. It catastrophically invaded the loess-covered Columbia Plateau in south-eastern Washington . . . and reached Pacific Ocean level via the Columbia River, 430 miles [694 km] or more from the glacial dam. So great a flood . . . has been estimated to have run for 2 weeks. It was 800 feet [244 m] deep through the Wallula Gap on the Oregon–Washington line.
>
> On the Columbia Plateau in Washington it transformed a dendritic preglacial drainage pattern into the amazing plexus of the Channeled Scabland. . . . It flooded across stream divides of the plateau, some of which stood 300–400 feet [90–122 m] above today's bounding valley bottoms. Closed basins as deep as 135 feet [41 m] were bitten out of the underlying basalt. Dozens of short-lived cataracts and cascades were born, the greatest of which left a recessional gorge, Upper Grand Coulee, 25 miles [40 km] long. The greatest cascade was 9 miles wide. The flood rolled boulders many feet in diameter for miles and, subsiding, left river bars now standing as mid-channel hills more than 100 feet [30.5 m] high . . .

So it goes on, with mentions of current ripples up to 3 m and more in height, a gravel delta 500 km² in area, the stripping off of the loess cover over an area of almost 5000 km² and so on. What is more, it seems that there was not just one bursting of the dam, but as many as seven, with the dam being re-implaced every time by fresh glacial advances. The cumulative result of all this was features such as Devil's Canyon, accommodating merely 'a distributary of a distributary', but cut 120 m into solid basalt.

At the end of a visit by a meeting of the International Association of Quaternary Research to the region in 1965, a telegram was sent to Professor Bretz which concluded: 'We are now all catastrophists'.

However, I would not wish it to be thought that this was necessarily a unique example. Even in the same region of the western United States there was the catastrophic breaking of the morainic dam of 'Lake Bonneville' (ancestor of the Great Salt Lake) about 30 000 years ago. This flooded the wide Snake River Plateau in Idaho with similar effects. Around the world generally one finds similar examples, albeit on less than an American scale, and there are probably many more not yet known in the world's literature. One reads of 52 m of debris being deposited in an hour as the result of a cloud-burst. One sees huge deposits, such as the high cliffs of gravel around Embrun in the French Alps, and not far away, the heap of great boulders at Claps de Luc, where half a mountain fell into the valley of the Drôme one wet afternoon in 1442 (Figure 5.2). The largest landslide of all in the Alps, that of Flims in Switzerland, is calculated to have brought down a mass of more than 12 km^3 of material. We know, therefore, that the frequency of landslides is quite enough to account for a major part of the wearing down of new mountain chains.

Particularly disastrous, but not uncommon, have been the effects of landslips and rock-falls into bodies of water. In 1958,

Figure 5.2 *Fifteenth-century landslip at Claps de Luc (Drôme) in the French Alps (DVA)*

40 million cubic meters of rock fell into Lituya Bay on the coast of Alaska, producing a great surge which destroyed a forest and reached more than 500 m up the mountainside on the far side of the bay. In 1792, a similar rock-fall into Shimbara Bay, on the Japanese island of Kyushu, caused three surges which drowned 15 000 people. The most spectacular known to me in Europe were the great rock-falls that occurred from the sheer face of Ramnefjell (Raven Mountain) into Loenfjord in central Norway in 1905 and 1936, producing waves that wiped out local communities and carried a steamer a considerable distance inland.

Most countries of the world have their records of great natural catastrophes which changed the local face of the earth. One thinks of the change in the course of the Hwang-ho River in China, which in less than 80 years moved its mouth some 400 km from way to the south of Shantung on the Yellow Sea up nearly to Tientsin on the Gulf of Pohai. The Brazos River in Texas is said to change its course abruptly once every 10 years or so.

Alec Smith has told me of his studies on the bottom sediment of Lake Windermere in north-west England. The rush of visitors to the area since the popular romanticism of the 'Lakeland poets' has led to major changes in the micro-organisms and the deposition of their remains, due largely to the effects of human effluent. He has therefore aptly called the higher layers the 'post-Wordsworthian'. This shows how rapidly a new fauna and/or flora can migrate into an area and change the record albeit on a very small scale.

Volcanic effects have been even more sudden and disastrous, ranging from the explosion of the island of Krakatoa, between Java and Sumatra, in 1883 to the even more catastrophic eruption of Santorini (or Thira) in the Aegaean about 1470 BC. This eruption, or series of eruptions, which resulted in the huge collapsed caldera in the sea beside the present island, must have been the greatest catastrophe ever witnessed by man and may well have been heard as far away as Britain. The whole eruption probably lasted only about 100 days, but was probably indirectly responsible for the destruction of the Minoan civilisation on Crete, nearly 100 km away. This seemed to fit in well with Plato's account of the end of Atlantis. The correlation of the eruption and the destruction was long disputed, but now seems to be confirmed. Nevertheless volcanic bombs were hurled this distance compared with a mere 40 km from the more publicised Krakatoa eruption. It is, however, still very easy to pick up tufa from the latter on the coast of eastern Africa, to which it took 6 months to float across the Indian Ocean.

One of the most spectacular sights ever seen by man must have been the mile-high (1.6 km) fiery cascade when a lava flow poured into the Grand Canyon in Arizona. Earlier lava flows, before the coming of man, date back a million years, but since that time the Colorado River has only cut down about 15 m. The canyon itself, more than 1½ km deep, cannot have started more than 10 million years ago, so here too there must have been some very rapid erosion at some time.

So we come back again and again to the notion of the rare catastrophic happenings playing a major role in the working out of the stratigraphic record as we find it today. Examples of this sort are in direct contrast to what has been, in effect, the subconscious attitude of most geologists for the last 100 or more years. The opposing attitude was perhaps best expressed by the great French naturalist, the Comte de Buffon, back in the eighteenth century. In 1781, he wrote:

> We ought not to be affected by causes which seldom act, and whose action is always sudden and violent. These have no place in the ordinary course of nature. But operations uniformly repeated, motions which succeed one another without interruption, are the causes which alone ought to be the foundation of our reasoning.

This may be a reasonable philosophical argument when we are thinking in brief human terms, but the stratigraphical record and I both seem to prefer the doctrine of Thomas Osbert Mordaunt: 'One crowded hour of glorious life is worth an age without a name.'

The hurricane, the flood or the tsunami may do more in an hour or a day than the ordinary processes of nature has achieved in a thousand years. Given all the millennia we have to play with in the stratigraphical record, we can expect our periodic catastrophes to do all the work we want of them.

Peter Gretener has emphasised the need to distinguish between *continuous* and *discontinuous* processes and I would like to adapt his list of examples as follows, ranging from biological to astronomical processes:

Continuous	*Discontinuous*
Phyletic gradualism	Punctuated equilibria
Erosion by a meandering river	Flash flood
Compaction of sediments	Internal collapse of sediments
Reef growth	Storm beach
Subsidence	Landslip
Diapirism	Faulting
Uplift	Earthquake
Pelagic deposition	Turbidity currents
Heat flow	Intrusions
Sea-floor spreading	Continental collision
Magnetic field	Magnetic reversal
Cosmic rays	Meteoritic impact

So the contrasting processes range through every aspect of what we call geology. Gretener went on to suggest a scale, based on a 95% probability of a particular event happening, as follows:

	Periodicity in years
Regular events	10^2, i.e. once in a human lifetime
Common events	10^3, i.e. once in human history
Recurrent events	10^6, i.e. once in the smallest stratigraphical unit
Occasional events	10^8, i.e. once in an era
Rare events	10^9, i.e. once in the history of the earth

Of course it does not matter what we call them, but this merely illustrates how the rare event merges into the regular. We must not dismiss the occasional and rare events as wild speculation; statistically they must occur sooner or later, just as sooner or later that chimpanzee will type *Hamlet*. All the time we must ask ourselves: 'Is the Present a long enough key to penetrate the deep lock of the Past?'

It is particularly instructive to look at the stratigraphical record of our kindred science of archaeology. This is close enough to us in time to qualify for the 'present' end of the uniformitarian doctrine. Buffon's 'operations uniformly repeated', the operations, 'which succeed one another without interruption', these are the ploughing, sowing and reaping, the building and decay of habitations, the births, marriages and deaths of human history. But they make very little showing in the archaeological record. It is the floods and the fires, the battles and the bombardments, the eruptions and the earthquakes which have preserved so much of the human story. When the palace is allowed to decay and the stones are taken away to build the shanty town, then the frescoes and the tesselated pavements, the statues and the beautiful pottery are all lost. It is only when the barbarian reduces the palace to a heap of stones in the desert and slaughters all the inhabitants, that the record of art and thought and everyday life is preserved for us. Think of that most perfect of all archaeological records—the city of Pompeii—where we have everything preserved from the statue in the elegant garden to the beans in the cooking pot, from the political graffiti on the walls to the obscene paintings in the brothel. These were all preserved by a single, brief catastrophe—the eruption of Vesuvius on the 24 August AD 79.

One interesting geological effect of the Vesuvius eruption nineteen hundred years ago was on the famous pillars of the so-called 'Temple of Serapis' at Pozzuoli on the opposite side of the Bay of Naples (Figure 5.3). These have long been associated with the archpriest of uniformitarianism, Charles Lyell, who used them as the frontispiece of his great proselytising work, *Principles of Geology*. The pillars show borings made by marine organisms several metres above the present highest sea-level and so provide undeniable evidence of marine invasions and retreats within historic times. These changes have been related to volcanic activity in this eruptive neighbourhood. They rose sharply, for example, during the eruption of Monte Nuevo in 1538. In fact what strikes me most forcibly about Lyell's pillars is not their evidence of placid uniformitarianism but rather of episodic 'catastrophism'. The borings are concentrated high on the pillars but are remarkably lacking on the masonry lower down. A gradual advance or retreat of the sea from its highest point would surely have produced pock marks all the way down to present sea-level. Instead they show to my prejudiced eyes that the sea changes were very rapid indeed.

Figure 5.3 *Lyell's pillars at Pozzuoli near Naples showing borings made by marine organisms during a marine invasion in post-Roman times (DVA)*

However, I would not for one moment deny the continuity and the gradualness of the processes which are changing the earth. But we must always distinguish between the nature of the process and the nature of the record. I do not deny uniformitarianism in its true sense, that is to say, of interpreting the past by means of the processes that we see going on at the present day, so long as we remember that the periodic catastrophe (including sudden events like the rush of a turbidity current) is one of those processes. All that I am saying is that I strongly suspect that those periodic catastrophes make more showing in the stratigraphical record than we have hitherto assumed.

That brings me then to my fifth proposition: *the periodic catastrophic event may have more effect than vast periods of gradual evolution*. This may be called the 'Phenomenon of Quantum Sedimentation'.

REFERENCES

Ager, D. V. (1989). 'Lyell's pillars and uniformitarianism', *J. Geol. Soc. London*, Vol. 146, pp. 603–605.
Showing how the 'Icons' of uniformitarianism (as they have been called) probably represent very rapid changes in sea level.
Boekschoten, G. J. (1971). 'Quaternary Tephra on Crete and the Eruptions of the Santorini Volcano', *Opera Botanica*, No. 30, pp. 40–48.
A summary of the evidence of the catastrophic destruction of the Minoan civilisation.
Bretz, J. H. (1923). 'The Channeled Scabland of the Columbia Plateau', *J. Geol.*, Vol. 31, pp. 617–649.
One of the first papers by this remarkable man on the sudden bursting of a morainic dam in Montana and its catastrophic effects.
Bretz, J. H. (1969). 'The Lake Missoula Floods and the Channeled Scabland', *J. Geol.*, Vol. 77, pp. 503–543.
A definitive paper, written almost 50 years later after much controversy, using the evidence now available from new road-cuts, man-made dams, aerial and satellite photographs and new detailed maps, to prove beyond dispute his original catastrophical hypothesis.
Coleman, P. J. (1968). 'Tsunamis as Geological Agents', *J. Geol. Soc. Australia*, Vol. 15, pp. 267–273.
A brief survey of tsunamis and their probable importance as neglected geological agents.
Curray, J. R. (1965). 'Late Quaternary History of the Continental Shelves of the United States', in H. E. Wright and D. G. Frey (eds), *The Quaternary of the United States*. Princeton University Press, Princeton, NJ, pp. 723–735.
A very interesting article in a very valuable source-book.

Ericson, D. B. and Wollin G. (1971). *The Ever-Changing Sea*. Paladin, London
A wonderfully readable source-book of fascinating information about present-day marine phenomena.

Evans, G. (1965). 'Intertidal Flat Sediments and Their Environments of Deposition in the Wash', *Quart. J. Geol. Soc. Lond.*, Vol. 121, pp. 209–245.
A detailed account of this interesting area of modern sedimentation.

Gretener, P. E. (1967). 'Significance of the Rare Event in Geology', *Bull. Am., Ass. Pet. Geol.*, Vol. 51, pp. 2197–2206.
Presents the thesis that given the vastness of geological time, the improbable is certain to have happened many times.

Hayes, M. O. (1967). 'Hurricanes as Geological Agents: Case Studies of Hurricane Carla, 1961, and Cindy, 1963', *Bur. Econ. Geol. Univ. Texas, Rept. Investigation*, No. 61.
Details of the catastrophic effect of hurricanes on topography and sedimentology and puts forward the hypothesis of quantum sedimentation.

Holmes, A. (1965). *Principles of Physical Geology* (2nd Edition). Nelson, London.
The great source-book of all the original thought in geology. If (as in this book) you think you have a new idea, you will almost certainly find it here already.

Malde, H. E. (1968). 'The Catastrophic Late Pleistocene Bonneville Flood, Snake River Plain, Idaho', *US Geol. Survey Professional Paper* No. 596.
The catastrophic breaking of a glacially-dammed lake, similar to that described by Bretz.

Marinatos, S. (1939). 'The Volcanic Destruction of Minoan Crete', *Antiquity*, Vol. 13, pp. 425–439.
The first suggestion that the eruption of the Santorini volcano brought to an end the 'golden age' of the Minoan civilisation on Crete.

6
The Process of Sedimentation

So far as the process of sedimentation is concerned, we need to consider two concepts which are fundamentally opposed to each other. These are what I may term 'the gentle rain from heaven' concept on the one hand, and 'the moving finger writes' concept on the other.*

The former is that more or less subconscious attitude to sedimentation which still seems to be held by many (if not most) stratigraphers, though not by students of Recent sediments. It is the presumption that during any particular moment in geological time, sediment was raining down everywhere, preserving all the different contemporaneous environments simultaneously. The resultant record of this 'gentle rain from heaven' process would be the traditional 'layer cake' stratigraphy, with each layer in this mixed metaphor representing several different facies.

The only place where this type of sedimentation seems to be going on at the present day is in the ocean depths, where the deposits consist mainly of the remains of minute pelagic organisms, literally raining down from a watery heaven, plus volcanic and meteoritic dust raining down more intermittently from the aerial heaven alone. It may also happen towards

---

*The quotations I might mention are from Shakespeare's *Merchant of Venice* and Fitzgerald's *Rubáiyát of Omar Khayyám*.

the edge of the continental shelves, but this is less certain. Even on the ocean floors, however, there are vast areas without sediment and great gaps within the sediment that is there.

On the main part of the continental shelves, which is the region that chiefly concerns us in the stratigraphy of the continents, the chief contribution of Recent sedimentary studies, in my opinion, has been the demonstration of lateral rather than vertical sedimentation. Modern deposits are not, it seems, laid down layer upon layer over a wide area. They start from a particular point and then build out sideways as in the traditional picture of a delta. In other words, all bedding is likely to be cross-bedding, though often on so gentle a scale as not to be recognisable in the field. It therefore follows that all sedimentary bodies, other than deep-sea oozes and volcanic ash deposits, are likely to be diachronous. This was also one of the main conclusions reached by Alan B. Shaw in his brilliant little book *Time in Stratigraphy*.

Such ideas would seem to contradict the lateral persistence of facies about which so much fuss was made in Chapter 1. But we will return to this point later. Let us consider a simple example from the stratigraphical column to see the implications of my conclusions.

In the Sirte Basin of eastern Libya, there are magnificent cliff sections, running literally for hundreds of kilometres, in what is usually referred to as the Marada Formation, of early Miocene (Burdigalian) age. Tectonic dips are negligible in this region and the winding escarpments, with many isolated 'jebels' or hill outliers in between (Figure 6.1) make possible a detailed investigation of the lateral variations in this so-called formation, on a scale beyond our dreams in more vegetated temperate terrains.

In effect, the scarp has been taken as representing the whole of the Marada unit, since it is commonly capped by a white limestone of post-Burdigalian age and the desert floor below is commonly strewn with *Lepidocyclina* from the Oligocene (sometimes neatly sorted by the desert wind into megalosphaeric and microsphaeric forms).

A sedimentary investigation of the area led to the postulation of five distinct facies within the Marada:

Figure 6.1 *Marada Formation (early Miocene) forming one of the 'Twin Buttes' in the Sirte Basin of eastern Libya, with an abandoned drill-bit symbolically in the foreground (DVA)*

    (i) an offshore sandbank facies;
   (ii) a lagoonal facies;
  (iii) an intertidal or deltaic facies;
  (iv) a fluviatile facies; and
   (v) an estuarine facies cutting across the other four.

The lagoonal facies has been disputed and the estuarine facies is complicated in various ways and confuses the issue, so let us just consider three of these facies: the offshore, the deltaic and the fluviatile. All three are very well represented as sediments, shelly fossils and trace-fossils. There is no dispute about this side of the interpretation, nor with the conclusion that the offshore deposits are mainly developed to the north and the fluviatile (continental) deposits to the south.

    There is disagreement, however, whether or not they can be interpreted as lateral time equivalents of one another. This is expressed diagrammatically in Figure 6.2. In Interpretation 1 (top) we have the 'gentle rain from heaven' approach, with all three

INTERPRETATION   1

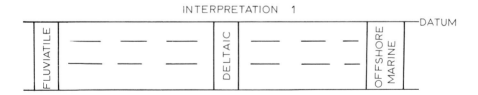

INTERPRETATION   2

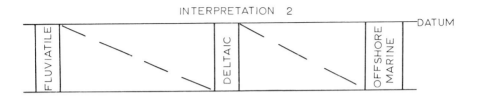

Figure 6.2 *Two possible interpretations of the relationships between the different facies of the Marada Formation, in the Miocene of the Sirte Basin of eastern Libya*

basic environments being preserved simultaneously. In Interpretation 2 (bottom) we have the 'moving finger writes' approach, which was basically the interpretation put forward by the palaeontologist who followed on in this particular research project. He found a certain amount of fossil evidence that the time planes were not parallel with the lithological boundaries and he came to the general conclusion that the facies to the north are in the main younger than those to the south. Clearly it is difficult to be dogmatic about faunas of this age, when such short time-spans are involved, and it is also difficult to be sure that particular species are more time-controlled than facies-controlled.

What is more, it is probably utterly unfair to this particular study to oversimplify it in the way shown in Figure 6.2, which is intended solely to relate to the limited area studied. Nevertheless, it does provide a simple demonstration of a problem that dogs all our stratigraphical thinking, and as the palaeontologist in question was my student (and I visited the area with him) my inclinations are wholly for the second interpretation in this case.

When stratigraphers discovered facies (way back in the days of Gressly), then all differences in lithology tended to become synchronous. Conversely, when they discovered diachronism, all similarities in lithology tended to be taken as evidence of different ages. In other words: 'if it looks different it must be of the same age', and 'if it looks the same it must be diachronous'. Obviously this is an overstatement, but it does not really exaggerate too much the state of mind many of us have reached in stratigraphical discussions. We have seen exactly the same psychological process in palaeontology, where the fashionable fixation for homoeomorphy in many groups brainwashed many of us into thinking: 'if they look alike they cannot be related'!

But if we accept the discoveries of the students of Recent sediments, we must accept the conclusion that Interpretation 2 in Figure 6.2 is more likely than Interpretation 1. Diachronism is, of course, a relative term and the degree of diachronism referred to here, within a single sedimentary basin, would be negligible if one were concerned with a much larger region. The important point is that by Interpretation 2 we mean that, *on the whole*, the deltaic deposits shown in the centre of the diagram were laid down later than the fluviatile deposits but earlier than the offshore deposits.

Going back to the apparent contradiction between this chapter and Chapter 1, we must consider this question of diachronism more carefully. As one of my examples of the lateral persistence of facies, I cited the mid Silurian limestones such as the Wenlock Limestone of western England. This does not mean that I am necessarily accepting a 'gentle rain from heaven' interpretation for the Wenlock Limestone and its equivalents. In this case I think it is more likely that limestone deposition started in several or many different centres and spread outwards. It may even be that at no one moment in mid Silurian times was limestone being deposited throughout the region concerned.

If I use the analogy of women's fashion, it is not that all the women in the western world suddenly decided one morning to cut a foot or more off all their dresses and to appear to a shocked and/or delighted male world in mini-skirts. It is also, thankfully, not true that they all decided simultaneously to allow their dresses to droop drearily down once more. The mini-skirt spread through the western world from many centres; it appeared in Oxford Street, London, long before it reached Oxford Street, Swansea, and presumably it was seen in Chicago, Illinois, an appreciable time before it dazzled the male eyes of the 'city' of Muddy in the south of the same state. In other words it was diachronous. Nevertheless, in the late 1960s, mini-skirts were distributed all over the world and could be (if preserved) the index fossil for that epoch. The same could be said of carbonate deposition in the mid Silurian.

Diachronism is not a phenomenon very much considered by Recent sedimentologists, simply because they have not got time enough to recognise it. Nevertheless it does occur, as has been clearly demonstrated, for example, in the intertidal and supratidal deposits of the Trucial Coast on the south-west side of the Persian Gulf.

All along the coast here, an algal mat is developed in the intertidal zone. This is formed mainly of blue-green algae and makes an efficient sediment trap. Similar deposits have been recognised in ancient sediments. But if one walks inland across the salt marsh or *sabkha* and digs a hole, one can find the same algal mat buried beneath wind-blown sand with layers of anhydrite and gypsum. In fact there is a direct continuation inland for several miles of the algal mat that can be seen forming on the beach at the present day. Samples taken of this carbonaceous layer more than 6 km inland were dated by their carbon-14 content and proved to be about 4000 years old (Figure 6.3). In other words, 4000 years of diachronism in 6.5 km. What could be more impressive? How slow we stratigraphers would seem to be in recognising the importance of such diachronism! And yet, is it so impressive? Four thousand years in 6.5 km is forty thousand years in 65 km, four hundred thousand years in 650 km and four million years in 6500 km. We are still dealing with almost negligible figures in geological terms and certainly of the same order as the synchronous/

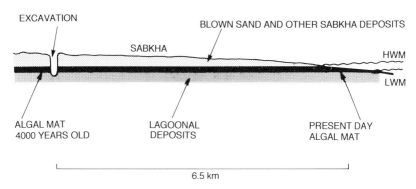

Figure 6.3   *Diachronism of an algal mat deposit on the Trucial Coast (from information in Evans and Bush, 1969)*

diachronous deposits such as the 'Urgonian' limestones discussed in Chapter 1.

It all depends on whether you think in human terms or in geological terms. The algal mat deposit is strongly diachronous to us ephemeral human beings, but even if it extended for hundreds of kilometres inland it would still be virtually synchronous geologically.

Let us consider a larger modern parallel: the Po Valley in northern Italy and the Adriatic Sea (Figure 6.4). We have here a wide, flat valley filled with sediment passing into a long parallel-sided sea reminiscent of many ancient sedimentary troughs. The Po delta has been moving forward irresistibly for aeon upon aeon since late Tertiary times and will presumably, in the course of time, fill the Adriatic with its sediments from the decaying Alps. The sedimentologist of the distant future will find a long trough filled with deltaic and fluviatile sediments with clear evidence of their provenance from the north-west and of longitudinal infilling like so many other trough-shaped basins up and down the stratigraphical column. To him the deposits of the migrating Po drainage system will appear to be synchronous. Yet to mere man, Venice stands today on a group of islands at the head of the Adriatic very much as it was at the time of the great Doges.

This brings us again to the vital question of where sediments actually accumulate at the present day. This was discussed at the beginning of the previous chapter, but it needs to be stated again that there seem to be comparatively few and small areas on the shelves at the present day where sediment is actively

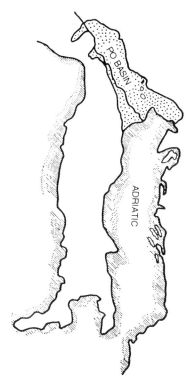

Figure 6.4   *Relationship of the Po Basin to the Adriatic Sea*

accumulating. There are plenty of areas of sea-floor with Recent sedimentary cover of sorts, but—at least on the inner shelves— this nearly always seems to be moving to and fro and not building up. Even in such classic areas as the Mississippi delta, where sediment is thought to be accumulating rapidly, there is plenty of evidence to suggest that, after building up for a while, much of it is carried away again. No doubt a great deal eventually comes to rest in the deep ocean basins, but these are not environments much represented in the stratigraphical record of the continents, which have been our main preoccupation. In fact, even before the days of plate tectonics, I have always been struck by the paucity of oceanic sediments in the continental areas. We can get rid of much of it by subduction, but certain orogenic episodes (notably the Variscan) seem to have very little to show of the ocean floor.

If we escape from the notion of sediment raining down everywhere, all the time, we also escape from the notion of the sea-floor subsiding all the time. One of the most important corollaries of the hypothesis of the turbidity current was that we could now have the geosynclinal trough without the geosynclinal sediments. This was clearly demonstrated in the Alps in what were half humorously called 'leptogeosynclines', that is troughs with very little sediment.

More important was the classic work on what is probably our best contemporary geosynclinal area—the Indonesian Archipelago. Here a great trough that is filled with Caenozoic sediments in northern Java and Sumatra (where a sediment supply was available) passes directly into a deep oceanic trough—the famous Flores Deep—where there was no such supply.

Obviously a great thickness of sediment must be heavy and must press down that part of the crust on which it rests, just as the weight of continental ice caused subsidence of the great land masses during the Pleistocene. But snow and ice accumulate in a totally different way from sediment. They do not require basins, in fact they prefer mountains. What is more, they operate on a completely different scale. The old-fashioned concept of the sedimentation causing the subsidence to accommodate it is just not tenable. It is a process of diminishing returns, as Arthur Holmes showed mathematically some 50 years ago. Obviously a given thickness of sediment could not simply by weight produce the same amount of subsidence of a much denser crust. In practice, for a given area to remain in isostatic equilibrium, there would always be a limiting factor. Holmes calculated a ratio of approximately 2.4:1. That is to say, for 100 m of water one could only expect 240 m of sediment of *normal density*.

Sedimentation must occur preferentially in certain areas, such as deltas, where there is a plentiful supply of sediment and a suitable retardation of the transporting medium. This explains the abundance of deltaic sediments in the stratigraphical record of the continental areas, but one must also expect that such sediments will either not exceed a critical maximum thickness, such as that suggested above, or they must have been deposited in a tectonically subsiding trough.

We know that the process of thick sedimentation followed by isostatic readjustment has happened frequently in the past, both in truly geosynclinal areas and in Voigt's 'bordering troughs' (*Randtröge*) of the shelf regions. But here tectonic pressures must have produced the subsidence and prevented the immediate re-establishment of isostatic equilibrium. In these cases, therefore, sedimentation was not the cause of the subsidence, but, as always, subsidence was the cause of the preservation of the sediments.

At times, of course, we know that the rate of subsidence (and the rate of uplift) has influenced the type of sedimentation, so the two are connected, but my general thesis remains that for the preservation of the bulk of the continental stratigraphical record we must think of the two as separate and independent phenomena. Sedimentation goes on all the time, for ever moving from place to place, for ever cannibalising itself. Subsidence—on the scale we are concerned with here—is generally a quite different matter and must be involved with the internal processes of the earth. It is only when sedimentation and subsidence coincide that the conditions will be right for the preservation of the vast thicknesses that constitute the stratigraphical record.

The conclusions we reach in this chapter, therefore, are that: *most sedimentation in the continental areas is lateral rather than vertical and is not necessarily directly connected with subsidence*. This may be called the 'Principle of the Relative Independence of Sedimentation and Subsidence'.

REFERENCES

Doust, H. (1968). 'Palaeonvironment Studies in the Miocene (Libya, Australia). Vol. 1, Zelten area, Libya. Unpublished PhD thesis, University of London, pp. 1–254.
A palaeontological and palaeoecological approach to the same sediments as those discussed by Selley (see below).

Evans, G. and Bush, P. (1969). 'Some Oceanographical and Sedimentological Observations on a Persian Gulf Lagoon', *Mem. Simp. Internat. Lagunas Costeras. U.N.A.M. U.N.E.S.C.O., Mexico (1967)*, pp. 155–170.
A useful summary of some interesting modern sediments with evidence of diachronism in the last 4000 years.

Selley, R. C. (1969). 'Near-shore Marine and Continental Sediments of the Sirte Basin, Libya', *Quart. J. Geol. Soc. Lond.*, Vol. 124, pp. 419–460.
A lucid survey of sedimentation along a Miocene shoreline.

Shaw, A. B. (1964). *Time in Stratigraphy*. McGraw-Hill, New York.
One of the most original books ever written in the field of stratigraphy.

Trümpy, R. (1960). 'Paleotectonic Evolution of the Central and Western Alps', *Bull. Geol. Soc. Amer.*, Vol. 71, pp. 843–908.
A survey of one of the classic geosynclinal areas of the world.

Voigt, E. (1963). Über Randtröge von Schollenrandern und ihre Bedeutung im Gebiet der Mitteleuropäischen Senke und angrenzender Gebiete', *Zeitschr. Deutsch Geol. Gesell.*, Vol. 114, pp. 378–418.
One of the most important papers published on European geology in recent years.

7
Marxist Stratigraphy and the Golden Spike

A few years ago, at a symposium in eastern Europe, I was chided for my non-Marxist attitude on stratigraphical theory. More recently (and more light-heartedly) it was pointed out to me after a lecture in England, that my ideas on the stratigraphical column were essentially Marxist in ideology. It seems that you just cannot win.

It may be asked how the great bearded father figure comes into the matter. The answer is that it depends whether or not you think that the history of the earth is divisible into units by means of natural events (or revolutions) detectable by man. The alternative is a record without natural breaks, only divisible by arbitrary man-made decisions. It is, if you like, dogmatism versus pragmatism.

The particular argument in which I was involved in eastern Europe concerned the base of the Upper Jurassic. W. J. Arkell, king of the Jurassic, had placed it in 1933 at the base of the Callovian Stage. This had been accepted as the 'party line' over much of the world. But in 1956 Arkell changed his mind and (partly, one suspects, for the sake of tidiness) pushed the Callovian down into the Middle Jurassic. This move considerably upset the Russians since, apart from anything else, it meant redrawing a lot of their maps. As the only Englishman present at the conference in question, I was called upon to defend the late Dr Arkell's change of usage.

The chief argument for putting the Callovian in the Upper Jurassic was that this stage was markedly transgressive over a large part of the Soviet Union and that, therefore, this was a natural break such as one might expect at a major stratigraphical boundary. At least, one would expect it if one was imbued with that particular philosophy. This therefore implies a sort of traumatic stratigraphy with major events, such as transgressions, virtually synchronous on a continent-wide or even a world-wide scale.

It may be argued that this is essentially the approach that I used in the first chapter. But now I am going to attempt a volte-face. The object of the first chapter was to draw attention to the basic oddity of the stratigraphical record in that particular facies were remarkably widespread during particular periods of geological time. It was said, in effect, that the extent of, say, carbonate deposits over a few square kilometres in the Bahamas at the present day is something quite different from the persistence of carbonates in Early Carboniferous times over much of the northern hemisphere. This is not to say that carbonate deposition began everywhere simultaneously as some heavenly clock chimed in the beginning of the Carboniferous Period. It is a problem not easily solved by the classic methods of stratigraphical palaeontology, as obviously we will land ourselves immediately in an impossible circular argument if we say, firstly that a particular lithology is synchronous on the evidence of its fossils, and secondly that the fossils are synchronous on the evidence of the lithology.

One can still argue, as I have argued myself in connection with the correlation of the north-west European Trias, that major events, such as marine transgressions on to one part of a continent, are likely to have more widespread effects in the rest of that continent. The British Trias is almost completely lacking in marine fossils below the Rhaetian, but I maintained that one could nevertheless see the effects of alpine transgressions reflected in our continental sediments. However, we are always on dangerous grounds if we accept this as anything other than a last resort in the absence of really adequate evidence of evolving fossil lineages.

We have managed to confuse ourselves for years with the jargon of lithostratigraphy, biostratigraphy, chronostratigraphy and the rest. In fact it can well be argued that basically there are only two concepts—rocks and time—with the rest just an

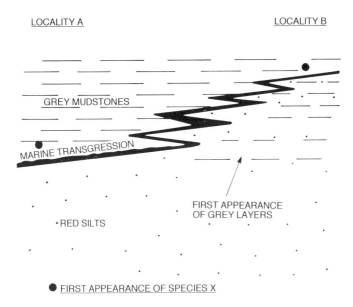

LOCALITY A   LOCALITY B

GREY MUDSTONES

MARINE TRANSGRESSION

FIRST APPEARANCE
OF GREY LAYERS

•RED SILTS

● FIRST APPEARANCE OF SPECIES X

Figure 7.1   *To illustrate the situation where 'event' correlation may be more accurate than palaeontological or lithological correlation, as in the Trias of north-west Europe*

obfuscation of the nomenclature. Nevertheless, it is useful to distinguish between our various means of correlation and I make no apology for suggesting another term, just to draw attention to its usefulness as a method. This is what may be called 'event stratigraphy'* in which we correlate not the rocks themselves, on their intrinsic petrological characters, nor the fossils, but the events such as the Triassic transgressions just discussed. The term has since caught on.

Figure 7.1 illustrates the very obvious way in which a new fossil species may seem to be diachronous. Such diachroneity may be very real and much more subtly controlled than this. Equally clearly, the main lithological characters may be similarly diachronous. What is less obvious is that the marine transgression, or whatever other event it may be at Locality A, could well be reflected in some way in the different facies at

---

*My friend John Gould (formerly Professor of Classics at Swansea) advised me that 'genomenostratigraphy' would be the most obvious term of Greek origin, but neither he nor I would wish to inflict such a monstrosity on an already over-wordy subject. Another possibility would be 'pathostratigraphy', but this would be misunderstood.

Locality B. In such a case (which must be, theoretically at least, very common), 'event stratigraphy' is more accurate, in a chronological sense, than either lithostratigraphy (in the usual sense) or biostratigraphy.

Over a period of many years I had a succession of research students working on the Eocene strata of the Isle of Wight, off the south coast of England. No doubt if their theses eventually

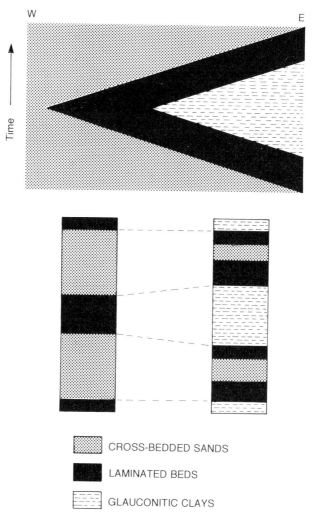

Figure 7.2   *Cyclic sedimentation and 'event' correlation in the Eocene of the Isle of Wight, southern England*

appear they will contradict my oversimplified ideas on the matter, but this seems to me an ideal place to demonstrate 'event stratigraphy'. Cyclic sedimentation has long been recognised here, with the succession at the east end of the island largely marine, and that at the west end largely continental. Correlation by fossils is, for the most part, impossible. Correlation by lithology always leads one into an impossible tangle.

There are basically three kinds of sediment involved (Figure 7.2):

(i) cross-bedded yellow and buff sands;
(ii) laminated beds of alternating layers of clay (or lignite) and sand; and
(iii) glauconitic or sandy clays with marine fossils.

If one tries to correlate one of these (say the laminated beds) from one end of the island to the other, one finds that there are just too many units at the east end. The paradoxical solution, it seems to me, is to correlate the unlike lithologies as shown in Figure 7.2. The glauconitic clays at the east end should therefore be equated with the laminated beds at the west end. It may be expressed as a correlation of the degrees of 'marine-ness'. We are then correlating equivalent points on the cycle. Anyway, it seems to work! This has, of course, been done before elsewhere, but the principle is not, I believe, sufficiently appreciated.

Undoubtedly, comparatively sudden and very widespread events, such as major marine transgressions, did occur at various times during the earth's history. The most famous of these is the Cenomanian transgression, exemplified by Figure 3.1, where Late Cretaceous sediments rest, with marked unconformity, on the Precambrian rocks of the Bohemian Massif. To call it the 'Cenomanian transgression' is something of an oversimplifica-tion, for it is often Albian or Turonian in age, but a major transgression at about the beginning of Late Cretaceous times seems to have occurred almost all over the world. It presumably represents an epoch of geomorphological maturity and plate stability.

Lithological continuity or marine transgressions or orogenic phases or any other physical phenomenon, must always be measured (in the absence of anything better) against the scale provided by organic evolution. The point about the Urgonian limestones, say, is that we know that they are of *about* the same age throughout Europe *in spite* of the fact that fossil evidence shows them to have started and ended at different times in different places. In stratigraphy we are primarily concerned with the starts and the finishes, not with the monotonous middles.

A transgression is, by definition, transgressive. I tried to show in Chapter 4 that such events may have happened, on a geological time scale, very quickly indeed. Nevertheless, it is theoretically unsound to define what we want to be a synchronous horizon at a level that we suspect to be diachronous. We come therefore to the problem of defining the individual stratigraphical unit, which is the basis of nearly all our troubles.

Let us consider the concept of the stratigraphical unit in historical terms which can be more easily appreciated than the unimaginable vistas of geological time. Let us take as an example the Edwardian Era, well remembered by some still living today.

First let us consider how we define the Edwardian Era. Its logical beginning was the coming to the throne of that 'comfortably disreputable' monarch, Edward VII on the 22 January 1901. The more tidy minded among us find it more convenient to think of the Edwardian Era as starting on the 1 January 1900, when his mother was still obviously, if obscurely, on the throne. The pedants would insist, however, that the 1 January 1901 was the true beginning of the twentieth century and therefore the obvious 'natural' marker point for a new era.

So we have three possible points for defining the beginning of the Edwardian Era. But defining the end of that era is even more difficult. Logically, I suppose, it was on the 6 May 1910, when Halley's Comet was making one of its occasional visits to our skies to blaze forth the death of princes, and King Edward VII joined his international ancestors. Indeed, though the death of a single organism may not seem to be very significant scientifically, it has been well argued by George Dangerfield that this one event did, in effect, mark the end or at least the beginning of the end, of that remarkable period of human history dominated by 'Liberal England'.

But most of us would probably say that the Edwardian Era ended with the bullets that Gavrilo Princip fired into the body of Archduke Ferdinand on the 28 June 1914. The more chauvinistic among us might place the marker in the calendar for 4 August of the same year when Great Britain entered the war against the central powers, or two years later at dawn on the 1 July 1916, when the opening of the Battle of the Somme destroyed the lives or the illusions of a generation of British youth.

Americans, who also used the term 'Edwardian Era', would probably say that it ended on 2 April 1917, when the United States entered the war. Many more (of varied political beliefs) would argue that the obvious dividing line in human history was the October Revolution which took place in Russia, in its paradoxical way, during November 1917.

So we see the difficulties of defining even a recent period of earth history, and the protagonists of the varied viewpoints would probably (if it mattered at all) argue just as fiercely as stratigraphers over the boundaries. But history is simple, it can be measured out by neat dates. Stratigraphical boundaries, with no real dates at all, are much more worthy and needful of vehement discussion, for these spell out our basic language.

Having used the Edwardian Era as an analogy in the definition of stratigraphical boundaries, let us consider the related problem of how we recognise it. Suppose no real dates were available, as with our older eras, then we would have to recognise it by the phenomena of organic evolution. We can quickly dispose of the most obvious zone fossil, the king himself, in that he had already lived for 60 years before his era began and we have seen that his era is commonly regarded as lasting several years after his extinction. What is more, the Edwardian Era is recognised and recognisable in many parts of the world where the 'zone fossil' never set foot.

We can perhaps recognise the era by other organic phenomena such as clothes, literature and music. But Elgar's pomposity and circumstance, and Galsworthy's saga are still very much with us and the 'Teddy Boy' style of dress reappeared in the 1950s as a sort of atavistic flashback. What were probably the most important characteristics of the era, the social and political attitudes, are not fossilisable at all. All these problems are exactly comparable with those we have further back down the stratigraphical column.

Many of the arguments over the Silurian/Devonian boundary, for example, relate to whether one's favourite fossils are fish,

graptolites, trilobites or brachiopods. I once heard a very distinguished palaeontologist argue that the base of the Devonian
was obviously at one particular level because he had shown that
one species of trilobite changed at that horizon into another.
Similarly with our historical pattern, veteran car enthusiasts
ignore all other cars, even the most regal of them, and define
'Edwardian' as the period from 1905 to 1918. For them the critical
starting date in human history was the one in 1896 when British
law was changed to allow motor-cars to be driven without a man
preceding them on foot carrying a red flag.

One of the most interesting papers on stratigraphical palaeontology published in recent years was one on the dating of old
mining camps in the American West by means of beer bottles
and beer cans. The old-fashioned bottle, shaped to take a cork
(Figure 7.3a) was replaced about 1900 by the bottle (still hand-
made) provided with a rim for a metal cap (Figure 7.3b). In the
1920s the bottles became machine-made and there was a progressive 'take-over' by the beer can with soldered edges and an
unfossilisable paper label (Figure 7.3c). During the 1930s

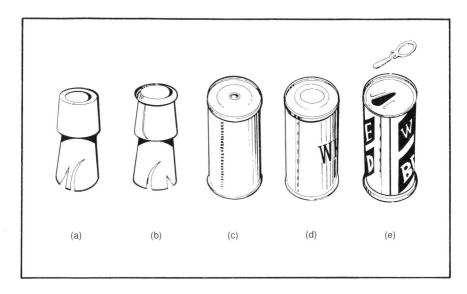

Figure 7.3 *Beer bottles and beer cans considered as stratigraphical palaeontology:*
(a) hand-made bottle for cork; (b) machine-made bottle for metal cap; (c) early tin can
with soldered joints and unpreserved paper label; (d) can sealed by crimping with label
printed on metal; (e) with tear-off metal flap ((a) to (d) after Hunt, 1959)

(presumably after the end of prohibition) this was replaced by the can sealed by crimping and with the label printed directly on the metal (Figure 7.3d). Since the above paper was published in 1959, a further evolutionary step was taken with the introduction in 1967 of the beer can with a tear-off flap (Figure 7.3e). Subsequently, in 1971, the circular ring started to be replaced by a D-shaped ring, since it was found that the former was being extensively used in parking meters! There was also another interesting evolutionary branch, not illustrated here, in which the early beer cans (like so many organic fossils) recapitulated their beer-bottle ancestry. In other words, many of the early beer cans imitated the bottle shape, although it was not a particularly efficient one for their material. (Many of us—old soldiers particularly—remember this primitive shape surviving in a form of evolutionary conservatism as the well-known and well-hated tins of 'Brasso' metal polish.) Beer cans of these various kinds have actually been used to date sediments presently being deposited off Baja California.

Besides being a beautiful illustration of straightforward stratigraphical palaeontology, the bottles and cans also demonstrate very clearly many of the other associated problems. Thus (like all other fossils) they are facies fossils. They are very abundant on the desert surfaces of the American south-west, but are presumably rare and localised in deep-sea deposits. They must be less common in wine-drinking regions and virtually unknown in strictly Islamic countries. They also display all the stratigraphical drawbacks of migration and diachronism. Thus the beer can evolved in North America in the 1920s and 1930s, but (if my memory serves me correctly) did not reach Europe until after the Second World War, except as erratic specimens transported by American servicemen.* What is more, they are now suffering severe competition as index fossils from the plastic bottle, which is (regrettably) much more easily preserved. In fact there is a humorous French classification of the topmost stratigraphical stage into:

---

*It has been pointed out to me that beer cans were in fact first made at Llanelli in 1935 during the depression of the 1930s, when they had nothing else to do with their tin-plate. I apologise—I was not then either a South Walian or a beer drinker.

> *Poubellien supérieur (à plastique)*
> *Poubellien inférieur (sans plastique)*

or in other words:

> Upper dustbinian/trashcanian (with plastic)
> Lower dustbinian/trashcanian (without plastic)

So we see that all the problems of correlation are just as real in recent history as they are in the stratigraphical record, once we have lost the advantage of the date at the top of the newspaper or letter. But still the principle of dependence on organic evolution proves the most satisfactory. Once the beer can had been invented, that is evolved, in one place, it was inevitable that it would eventually take the place of the bottle all over the world, though the process is still going on.

It is the startling and complete change over a large area that we have most to distrust. If the Palaeolithic stone axe is immediately succeeded in a section by a plastic bucket, then we must suspect a gap. We must not use this section to define our boundary or we will overlook not only the Edwardian Era but most of the rest of human history besides.

Yet many palaeontologists and stratigraphers still talk of defining boundaries at 'faunal breaks' as though there was a new creation at every stratigraphical boundary and the fossils above the boundary had no ancestors. This is all part of the attitude in stratigraphy that I may call 'the quest for the golden horizon'. This is the unspoken assumption which seems to underlie much stratigraphical thought and which says in effect that if one looks (and argues) long enough and hammers hard enough, then eventually one glorious day one will come upon the golden horizon that really *is* the Silurian/Devonian boundary. It is the assumption that the magic moment that was the beginning of the Devonian was ordained by God or Marx long before Man started his investigations.

The only real alternative to the 'golden horizon' is the 'golden spike'. This has none of the mysticism about it, but has been hammered in by a pragmatic human being, after careful choice of the most suitable section available. Many countries, especially on the European continent, still favour the 'stratotype' concept, whereby a stratigraphical division (normally a stage) is defined by reference to a type section or stratotype, at or near the locality mentioned in the stage name. This objective, though still sought as the panacea for all stratigraphical ills, has caused many of the problems that afflict us today. Thus the Bacchanalian and Machiavellian Stages, though theoretically adjacent in time, will inevitably be defined at their two different type localities. It is extremely unlikely that the top of the Bacchanalian at its type locality will exactly correspond with the base of the Machiavellian in its home ground. There may be an overlap, with resultant arguments between the protagonists of the two stages (like the classic dispute between Sedgwick and Murchison in the heroic age of geology).

Alternatively, strata will later be discovered that appear to fall into the time gap between the two stages. The resultant pseudo-scientific arguments will then concern themselves with the meaningless question as to whether the fauna of the intervening strata pertain more to the stage below or to the stage above. We are then back to the 'quest for the golden horizon' again, with the illogical presumption that there really is an answer to the question.

British Jurassic workers have been particularly disillusioned about the stratotype because, though Jurassic stratigraphy has always led the rest of the column, many of the classic Jurassic stages were derived from English place-names (like Kimmeridge and Bath) by a Frenchman (Alcide d'Orbigny) who never visited England.

The British Mesozoic Committee (who concerned themselves with such matters) therefore found it impossible to accept the stratotype concept as it is usually proclaimed on the continent. They chose instead to define only one boundary (the lower) of each division and to define it by a 'golden spike' (unfortunately only hypothetical) driven into the most suitable horizon in the most appropriate section.* This 'topless' fashion, as it has been called, has the immense advantage that the base of one

---

*The one serious critic of my first edition told me that this was in fact proposed earlier by the 'Ludlow Research Group' for the Silurian. However, that same group has since been busy revising the 'stratotypes' that I so much dislike, so I am unrepentant.

division then automatically defines the top of the division below. There can be no further arguments about gaps or overlaps. The base of the upper division can be defined as precisely as the lowest grain of sediment above the golden spike, so that even if there is a break at the level chosen, the definition will still stand and missing strata below the spike will automatically belong to the lower division.

This whole idea was subsequently taken up by the Stratigraphical Committee of the Geological Society of London and published, in a booklet, as their 'Provisional Code of Stratigraphical Nomenclature'. This chapter is, in effect, an unofficial part of our efforts to proselytise the idea around the world.* It immediately protects us from the impossible situation one meets in the literature with remarks such as: 'The Aalenian of Mr X, which is the Bajocian of Mr Y, should in fact be regarded as part of the Toarcian.' Not only are different names used for the same thing; different things are known by the same name. Thus certain of the Jurassic stage names used by some French workers do not even overlap in their usage between one specialist and another. There is also the so-called 'International Guide', edited by Hollis Hedberg (1972), although this is largely American in origin, and there are other versions in other countries.

Ideally, the golden spike (or 'marker point' as it is more prosaically called) should be chosen in a section where sedimentation seems to have been as nearly continuous as is ever possible, where there are no marked lithological changes and where there are unbroken records of several different groups of fossils. It is no good defining a chronostratigraphical boundary between, say, the Machiavellian and the Bacchanalian at the lithological junction of the Elgin Marbles and the George Sand. The faunas and/or floras of these two formations are likely to be different for purely ecological or preservational reasons (in fact the lower formation in this case, having been metamorphosed, will presumably have no recognisable fossils anyway).

---

*The first international decision to define a boundary in this way was taken at the 24th International Geological Congress in 1972 for the Silurian/Devonian junction. Appropriately, and onomatopoeically, the first 'golden spike' was hammered in at a locality in Czechoslovakia known as Klonk.

In other words, there will be a marked 'faunal break' and the boundary is automatically suspect.

It would be much better to choose, not the level where the genus *Euphoria* is succeeded suddenly by the totally unrelated genus *Amnesia* (as in Figure 3.8: 2 and 3 on left), but within a formation of uniform lithology where, for example, the species *Abra cadabra* (which really exists) passes insensibly into a descendent species or subspecies by a progressive statistical swing in the unit characters. If the chosen section has the greatest possible diversity and abundance of rapidly evolving fossils, we can then hope for the greatest possible number of different means for correlation.

Even if, through arguments of history, priority or convenience, we are persuaded to choose a section and a horizon that is less than perfect in the above characters, this still does not detract from the value of the golden spike as the ultimate arbiter for a particular boundary. We are not to know that new methods of correlation will not be developed (as spores, hystrichospheres, etc., have been developed in recent years) to correlate the least promising-looking formations. Ultimately perhaps we shall have a little black box into which we only have to pop our rock specimen for its age to be read automatically on a dial. Even then our marker point will have preserved for us a stability in stratigraphical nomenclature and will have saved us from the utterly wasteful vacillations in opinion and fashion that trouble us today.

Most of the talk around the world has been of 'stages', as though these units, though of no defined dimensions, are the ultimate in stratigraphical correlation on a world-wide scale. When a Devonian stage in Canada may be 3000 m thick and a Jurassic stage in Sicily may be thinner than its characteristic ammonites, we cannot altogether ignore sheer size. What is more, a Tertiary stage looks very much like a zone through Mesozoic eyes, and with Palaeozoic spectacles a Jurassic stage is at most an Ordovician zone. Also it must be said that Jurassic zones, which are the best known, are now being recognised on something approaching a world-wide scale.

In practice, therefore, we must go to the smallest convenient unit as our basis of correlation. The smaller unit must define the larger. Thus the base of the Epicene System must be defined at the bottom of its basal Binomial Series, and the base of that series must be defined at the bottom of the basal Para Zone. Ultimately we have the base of the Jurassic defined by a single bedding plane on the coast of Somerset, where we took the Jurassic specialists of the world to see it during our celebrations of the bicentenary of the birth of William Smith in 1969. (They have now been foolish enough to move it again on historic grounds!) Nevertheless, we still have international meetings planned to discuss this very matter. It does not matter whether the golden spike is hammered in somewhere in England or in France or in China, so long as we can make an arbitrary decision, stop arguing about words and get on with the much more difficult (but much more rewarding) task of correlation.

In this connection it is interesting to go back to William Smith, the father of stratigraphy, and to find him commenting in his memoir to the first geological map in 1815: 'The edges of the strata . . . are called their outcrops; and the under edge of every stratum, being the top of the next, and that being generally the best defined, is represented by the fullest part of each colour.' Of course, William Smith was defining lithostratigraphical rather than chronostratigraphical units, but the principle is the same.

Let us make an arbitrary decision (by a show of hands if necessary) *to define the base of every stratigraphical unit in a selected section*. This may be called the 'Principle of the Golden Spike'. Then stratigraphical nomenclature can be forgotten and we can get on with the real work of stratigraphy, which is correlation and interpretation.

REFERENCES

Ager, D. V. (1964). 'The British Mesozoic Committee', *Nature, Lond.*, Vol. 203, No. 4949, p. 1059.
In which the principle of the 'golden spike' is first introduced.
Ager, D. V. (1970). 'The Triassic System in Britain and Its Stratigraphical Nomenclature', *Quart. J. Geol. Soc.*, Vol. 126, pp. 3–17. A discussion of the correlation between continental and marine deposits based on the principle that a marine transgression in one part of a continent must have widespread effects elsewhere, especially in the centre of depositional basins.

Arkell, W. J. (1933). *The Jurassic System in Great Britain*. Clarendon Press, Oxford.

Still the great source-book on the British Jurassic.

Arkell, W. J. (1956). *Jurassic Geology of the World*. Oliver and Boyd, Edinburgh.

A monumental work which, along with everything else, redefined some of the basic units of the Jurassic and demonstrated the contradictions of many common usages.

George, T. N., Miller, T. G., Ager, D. V., Blow, W. H., Casey, R., Harland, W. B., Holland, C. H., Hughes, N. F., Kellaway, G. R., Kent, P. E., Ramsbottom, W. H. C. & Rhodes, F. H. T. (1967). 'Report of the Stratigraphical Code Sub-Committee', *Proc. Geol. Soc. Lond.*, No. 1638, pp. 75–87.

The principle of the base-marking 'golden spike' (here called a 'marker-point') included in the British Provisional Code on stratigraphical nomenclature.

George, T. N., Harland, W. B., Ager, D. V., Ball, H. W., Blow, W. H., Casey, R., Holland, C. H., Hughes, N. F., Kellaway, G. A., Kent, P. E., Ramsbottom, W. H. C., Stubblefield, J. & Woodland, A. W. (1969). 'Recommendations on Stratigraphical Usage', *Proc. Geol. Soc. Lond.*, No. 1656, pp. 139–166.

Complementary to the previous report with the principle of the 'golden spike' re-emphasised and applied to the definition of Jurassic stages and zones.

Harland, W. B., Ager, D. V., Ball, H. W., Bishop, W. W., Blow, W. H., Curry, D., Deer, W. A., George, T. N., Holland, C. H., Holmes, S. C. A., Hughes, N. F., Kent, P. E., Pitcher, W. S., Ramsbottom, W. H. C., Stubblefield, C. J., Wallace, P. & Woodland, A. W. (1972). 'A Concise Guide to Stratigraphical Procedure', *J. Geol. Soc. Lond.*, Vol. 128, pp. 295–305.

Hedburg, H. D. (ed.). (1972). 'Introduction to an International Guide to Stratigraphic Classification, Terminology and Usage', *Lethaia* (Oslo), Vol. 5, pp. 283–295.

Hunt, C. B. (1959). 'Dating of Mining Camps with Tin Cans and Bottles', *Geotimes*, Vol. 3, pp. 8–10, 34.

This is the original paper in which beer bottles and beer cans were first used in stratigraphical palaeontology.

McLaren, D. J. (1977). 'The Silurian–Devonian Boundary Committee. A Final Report', in A. Martinsson (ed.), *The Silurian–Devonian Boundary. Internat. Union Geol. Sci.*, Ser. A, No. 5, Schweizerbart. Verlags, Stuttgart, pp. 1–34.

The last word on Klonk.

8
The Nature of the Control

We come therefore to synthesise the ideas I had tried to put over in the preceding chapters. The main and very unsatisfying conclusion that I have reached may be expressed in the title I sometimes give to a lecture on the subject: 'There's something damn funny about the stratigraphical record.' The record is spasmodic and ridiculously incomplete, with particular strata and fossils extremely widespread, but separated by vastly longer gaps than anything that is preserved. The same strata and fossils, though to all intents and geological purposes synchronous, must have spread diachronously. Traditional concepts such as gentle, continuous sedimentation (and perhaps similarly continuous evolution) are not adequate to explain what we see. Nor is the concept of the 'stratotype' satisfactory as a means of establishing an international stratigraphical language. The record is spasmodic and must be treated as such. The 'layer cake' analogy just will not do. This applies equally whether we are talking about the sedimentary layers of the cake itself or the palaeontological currants within each layer.

We may consider rather the analogy of carpets being brought periodically into a shop for display and rolled out one by one on a pile. (I am thinking of the carpet shops of a few years ago. Nowadays they seem to display them all in a rolled-up condition, which spoils my analogy.) The resultant succession certainly looks like a 'layer cake', but the process of formation and the record it preserves are different. Superficially it is no more than a succession of parallel layers. But in fact we know that the time-gaps between successive layers may have been very considerable. We also know that when a new layer arrived, it was not deposited simultaneously all over the preceding layer; it was unrolled from one side or the other, so that the actual contact was progressive rather than synchronous. One might expand the metaphor further by pointing out that in the interval between the arrival of successive carpets, several of the existing pile would probably have been sold and removed from the succession. Also the height of the showroom (and the height of the prospective buyers) provide an ultimate control on the thickness of the pile.

This analogy is in every way applicable to my concept of the stratigraphical column. It differs from the more-or-less subconscious concept of 'the gentle rain from heaven' type of sedimentation, which probably only applies in special circumstances such as those of the oceanic oozes. It also differs from the 'layer cake' approach in not visualising layers which are parallel in both space and time. It recognises that sediment usually accumulates laterally rather than vertically and that almost every sedimentary body is therefore diachronous in human terms, though this diachronism is very rarely detectable in geological terms. Thus the time taken to unroll a carpet is very short compared with the time interval between the arrival of successive consignments.

In a sense this may be thought to contradict what I said in Chapter 1, where the emphasis was on the evident synchroneity of particular deposits, particularly carbonates. But again this is a matter of relativity. Thus the mid Silurian Wenlock/Gotland/Niagara limestones were obviously not completely synchronous. They did not begin forming everywhere simultaneously, and carbonate sedimentation did not cease throughout the northern hemisphere with the waving of some magic and/or deistic wand. More probably, carbonates grew outwards laterally from several (perhaps many) different centres, to fuse or overlap. But we still need some circumstance, or group of circumstances, to trigger off such accumulatory and, it must be emphasised again, episodic sedimentation.

My general view on this matter, as a suitably humble outside observer, is that the student of modern sediments pays too much attention to the way these sediments are laid down, their form and composition, but not enough attention to the question of whether or not they stand any chance of preservation for the stratigrapher of tomorrow. We think, perhaps, too much of the sedimentary environment, and not enough of what may be called the geophysical environment that can ensure their preservation.

A highly over-simplified picture of the way sediments accumulate for the future stratigrapher is expressed in strip cartoon form in Figure 8.1. At the top, in '1' we see the basic environmental situation with a landmass to the left, on the margin of which are found fluviatile, near-shore and off-shore sediments. The sediments are not building up vertically; they are in a constant (albeit commonly intermittent) state of change. Then in '2' we see a postulated tectonic situation with subsidence of a marginal belt. This results in a situation, represented in '3', where near-shore sediments can actually accumulate, presuming the subsidence to have occurred in that particular environmental belt. There is then a dynamic situation, with the near-shore sediments building out laterally in the usual way. We then arrive at '4' with the completion of a lithostratigraphical formation 'x' consisting of a seemingly uniform deposit of near-shore sediments, but—as I have already indicated—diachronous within itself. This may well be the isostatic limit discussed previously.

Now suppose we have a change in the tectonic situation, resulting, perhaps, from the stable situation reached in the near-shore area. All this time, it must be realised, the fluviatile, near-shore and off-shore environmental belts have remained more or less in the same position. The subsidence now takes place further out from the landmass, in the off-shore belt ('5') perhaps simply by the widening of the depression with continued subsidence. This results in a new accumulatory situation, shown in '6', with off-shore sediments now being preserved. Again lateral deposition will occur, but this time the sediments will have their diachroneity rather better disguised than that of the near-shore deposits. They may, in fact, come to rest on top of the earlier formation, at least at the edge. With the end of that accumulatory episode we arrive at the second resultant stratigraphical situation as suggested in '7'. We now have a stratigraphical set-up that has a second formation 'y' which seems to pass laterally into 'x' and to be its time equivalent.

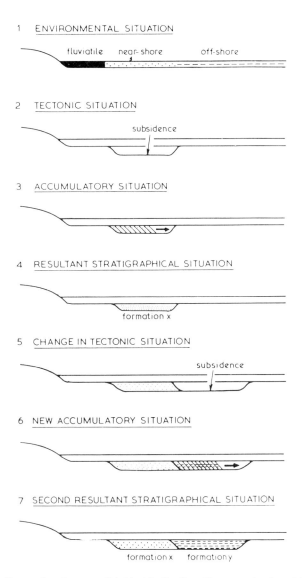

1 ENVIRONMENTAL SITUATION

fluviatile   near-shore   off-shore

2 TECTONIC SITUATION

subsidence

3 ACCUMULATORY SITUATION

4 RESULTANT STRATIGRAPHICAL SITUATION

formation x

5 CHANGE IN TECTONIC SITUATION

subsidence

6 NEW ACCUMULATORY SITUATION

7 SECOND RESULTANT STRATIGRAPHICAL SITUATION

formation x   formation y

Figure 8.1 *Successive diagrams (highly idealised) to illustrate the deposition of two formations which seem to be lateral equivalents, but in fact representing separate episodes diachronous within themselves*

From the stratigraphical point of view we have two formations. We know (from our god-like vision of their history) that 'y' is largely later than 'x' in time. We also know that each formation is diachronous within itself. But time-wise, the gap between them may well be much more important than the time-span within them. In other words '3' and '6' may be very brief episodes, whereas the pause between '4' and '5' may be very long.

Even more important, perhaps, is the realisation that all through this long history, the environmental belts have hardly changed their positions or their nature. The three environments were there all the time, just oscillating backwards and forwards a little, in relation to uplift inland. The environments did not change, it was the changes in the tectonic situation that ensured the preservation of the two formations. It must not be forgotten, of course, that subsidence itself may modify the nature of the sediment accumulated.

My example may seem a little strained and artificial but it is, I submit, a more realistic representation of the process of sedimentation (in the sense of accumulation) than the 'gentle rain from heaven' tacitly accepted by most of us. The 'moving finger' of the dynamic tectonic situation 'writes' the stratigraphical record for us, and only the erosional 'tears' accompanying subsequent uplift can wash out a word of it.

This then is the basic tectonic control of the stratigraphical record, in the sense of heaps of sediment being preserved for our study.

We come then inevitably to the question as to whether or not there is a pattern, cyclic or otherwise, in this record. In the preface to his great *History of Europe*, H. A. L. Fisher wrote: 'Men wiser than and more learned than I have discerned in history a plot, a rhythm, a predetermined pattern. These harmonies are concealed from me. I can see only one emergency following upon another as wave follows upon wave . . .' It seems to me that the same is true of the much older history of Europe, which is the subject of this book.

Of course, the historical analogy is not a completely valid one for history only concerns the interactions within a single species during a few thousand years, in connection with which geological and climatic changes have played only a minor role (as in the extinction of Carthage and of the Viking settlements in Greenland).

Many geologists have, like the allegedly wiser men mentioned by Fisher, claimed to see patterns in earth history. Thus that great American geologist, A. W. Grabau, wrote a whole series of books

entitled *Paleozoic Formations in the Light of the Pulsation Theory*.
More recently the redoubtable Larry Sloss has presented very
convincing arguments, from his very wide experience, of major
episodes of emergence and submergence, both on the North
American craton and on cratons generally.

Many more have recognised pulsations on a smaller scale,
especially in multiples of the magic figure seven or the sunspot
cycle of 11 years. The seasonal rhythm of varved deposits and
the daily rhythm of fossil coral growth are now well known.

However, questions of cyclicity produce much warmth and
much paper. I am fortunately not here concerned with the
controversies of cyclic sedimentation, but I am concerned with
the somewhat grander topic of orogenic cycles. There are any
number of 'mega' theories about the history of the earth. In
particular, there is the standard concept of the orogenic cycle.
Starting from the geosynclinal phase, there is the eugeosynclinal
trough filled with turbidites, greywackes or flysch (according to
where one is on earth and in the stratigraphical column), and
there is the Steinmann trinity of serpentines, spilitic pillow lavas
and chert (which we now are taught to call the 'ophiolite suite').
There is also the miogeosynclinal trough, with its volcanic-free
quartz sands and carbonates. Then comes the orogenic phase
with its folding and thrusting, its metamorphism and its granite
emplacement, passing into a phase of block faulting. Following
this we have the long period of isostatic emergence, rapid
weathering of the new mountains with the formation of molasse-
type deposits, widespread red bed sedimentation and also
perhaps a period of glaciation, before peneplanation, widespread
marine transgressions and the recommencement of the cycle in
new troughs. In the superficially simpler terms of plate tectonics,
these are the repeated effects of plates splitting and coming
together.

There is also the very obvious repeated control of certain forms
of sedimentation by climatic factors. Carbonates and evaporites
are obvious examples. Many authors have used the distribution
of particular carbonates to support particular theories of
continental distribution in the past. I do not presume to argue
with these, but our modern parallels do not help us overmuch.
It has been well argued, for example, that the sabkha cycles of
carbonate deposition are the equivalent, in arid tropical climates,

of the coal measure type cycles of the sub-tropical or temperate zones. Therefore the great extent of one or the other at various times in the past, might be no more than a measure of the width of the contemporary climatic belts. But it can hardly be argued that either carbonate or coal measure deposition is going on around the world today in anything like the way it has in the past. The nearest approach, perhaps, is the long belt of deltas extending today, from the Ganges in eastern India, via the Brahmaputra, the Irrawaddy and the Sittang, to the rivers of the Gulf of Siam and on via the Mekong to the Sang-koi and Si-kiang rivers of southern China. But even these are interrupted by the mountain chains of Burma and Malaya and only cover less than 15° of longitude.

Recently the coal measure type of cycle has been very plausibly explained in terms of climatically controlled ice-sheet surges, and we can very soon get lost in a multitude of explanations of the phenomena I have been discussing in this book. Two, however, deserve special mention in passing. One is the suggestion that periods of great carbonate deposition correlate with peaks in the productivity of volcanoes and result from the presence of more carbon dioxide in the atmosphere. This has been blamed, for example, for the extensive carbonates of the late Devonian and early Carboniferous, and for those of the Cretaceous. It has also been suggested that there has been a non-recurrent evolution of carbonate rocks, with a progressive diminution of carbon dioxide in the atmosphere since late Mesozoic times. It is also worth thinking about the possibility that rearrangements of the continents and oceans would themselves have considerable climatic effects. Thus the grandiose Russian proposal to melt the Arctic ice-cap by means of nuclear power may be criticised on the grounds that, since it would produce a new oceanic area (with increased peripheral precipitation) in high latitudes, this would result inevitably in another glaciation of Pleistocene proportions.

The second favoured whipping-boy, especially for widespread transgressions and regressions, has been the periodic orogeny. Many have suggested that the rise of a new median ridge within an oceanic basin might well be enough to flood great areas of continents. Thus the major cyclic changes in Jurassic sedimentation (and possibly even the changes of faunal and floral evolution) have been blamed on eustatic changes in sea-level, and these in turn on epeirogenic movements on the ocean floor. A more original suggestion, put forward recently, was that orogenic movements within the continental masses must cause eustatic *falls* in sea-level. The argument was that crustal shortening within the continents would produce the same displacement of underlying layers, but the seas would be spread wider and therefore lower. It was calculated that, assuming the average depth of the oceans to be about 4 km, then a 1% increase in the area of the oceans would be enough to lower the world-wide sea-level by about 40 m. This could easily be produced by orogenies occurring simultaneously in two or more continents. This theory was used to explain the major regressions of late Jurassic, late Cretaceous and late Tertiary times, with the major retreat at the end of the Mesozoic being blamed on the collision of India and Asia at about this time. In terms of plate tectonics it is easy to see how regressions can be explained by continental collisions, mountain building and the resultant displacement of water (Figure 8.2).

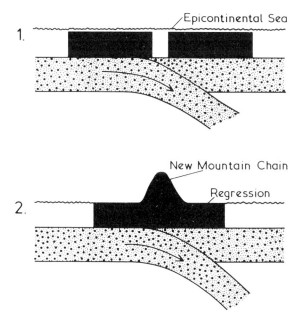

Figure 8.2 *Successive diagrams to show the displacement of water resulting from a continental collision. The same area of 'continent' and 'ocean' is present in each case*

Perhaps much else besides will be explainable in such terms. Professor Gilbert Kelling has suggested to me that the widespread early Cambrian and early Ordovician quartzites (discussed in Chapter 1) may represent episodes of plate stability. At such times, topographies will have become reduced in all the continents, erosion will have slowed up, and only very mature, multicycled sediments would have accumulated on the continental areas.

Now that we seem to be approaching, by way of plate tectonics and sea-floor spreading, some sort of general theory of earth history acceptable to the physicists, it is perhaps unwise to continue to hypothesise in this way. But I was trained in an era when we were told the continental drift was all right for the unscientific geologists, but the 'real' scientists—the physicists— said it was impossible. And I remember an earlier epoch when the geologists were told by the physicists, including the great Lord Kelvin himself, that they were foolish to postulate so many millions of years for earth history, when it could easily be calculated from classic physics that this was quite impossible. So I think we geologists should not be too bashful to theorise on the basis of purely geological evidence, and I cannot avoid the conclusion that at least some of the enigmas I have been discussing may have an origin in the climate.

It is fairly evident that the widely spread limestones and dolomites of Chapter 1 are, at least in part, explicable in terms of a wider tropical belt than we have at present. There is so much evidence of this that it hardly needs restatement here, nor does its corollary that the present tropical belt is atypically narrow. This phenomenon may also explain some of our anomalous fossil distributions and extinctions. Many fossil distributions, of very varied ages, imply remarkably uniform conditions over a wide part of the earth's surface. It has been suggested that the main controlling factor, so far as shallow marine organisms are concerned, is the 15°C winter isotherm, which corresponds pretty well with the present limit between the temperate and the sub-tropical marine faunas. This is also the most significant break in modern faunas. It has also been pointed out that there are many changes in the fundamental properties of water (and therefore presumably of protoplasm) at this level. On this basis, one can plot in the tropical belt for most periods of geological time, though the margin of error is such that sometimes they fit in with drifting hypotheses and sometimes they do not. Thus the archaeocyathids (or Pleospongia) of the Cambrian, certain brachiopods and forams in the Permian and the rudistid bivalves in the Cretaceous all seem to parallel the present equator. Other distributions, such as many of the Palaeozoic corals, evidently do not.

An interesting possibility is that, with a wider tropical belt than at present, one might expect an equatorial hyper-tropical belt around the equator, with minimum temperatures not known in modern seas. The winter isotherm for this might well be above the next critical temperature for water, i.e. 30°C. It would then be a reasonable deduction that some of the ancient equatorial groups that have since disappeared, such as the rudists, may have been adapted to these higher temperatures, and become extinct when they were no longer available. It hardly explains, however, the simultaneous extinction of the cooler water dwelling belemnites.

The poet's eyes (according to Shakespeare's Theseus) 'in a fine frenzy rolling, doth glance from heaven to earth, from earth to heaven'. So, ultimately, must the eye of the geologist, in seeking the nature of the control. One always seems to come back to climate as the primary explanation of the sort of phenomena I have been discussing, but for the ultimate control, sooner or later, we must face the possibility of an extra-terrestrial cause, though in most geological circles one seems to be expected to blush when doing so. I mentioned, in Chapter 2, the possibility of meteorites or variations in cosmic rays as a cause of abrupt changes in faunal successions. Such hypotheses have been postulated by highly reputable geologists when no other possible cause can be found to explain certain phenomena. I make no apology for joining a distinguished band of predecessors. Changes, cyclic or otherwise, within the solar system or within our galaxy, would seem to be the easy and incontrovertible solution for everything that I have found remarkable in the stratigraphical record. But perhaps we are not very far from finding such proof, if the stratigraphical record of the Moon or Mars proves to parallel in any way that of Earth.

However, I would not wish to end this book in an atmosphere of science fiction. If we cannot accept flying saucers, we must at least accept floating plates. It would be utterly inappropriate in 1992 to try to to consider the nature of the stratigraphical record, or indeed of any major aspect of geology, without seeking its relationship with the ideas of sea-floor spreading and plate tectonics. This is not just climbing on to a fashionable bandwagon; it is facing up to the fact that for the first time in the history of our science we are approaching a general theory of the earth. We are beginning to see how the whole thing fits together and again, I think, we are beginning to see a somewhat 'catastrophic' picture.

If we try to relate the stratigraphical column, in all its fantastic detail, to these general theories, we will inevitably soon bog down in a welter of information. I shall therefore only attempt to generalise about those parts of the world which I know best, and I shall be ruthless in the selection of my facts. That is not to say that I shall select my facts to fit my theory; I shall rather select those general impressions that I have gathered in the course of my geological wanderings. Inevitably it will be said that it is too soon for such syntheses. I am unrepentant; perhaps it will encourage someone else to do better.

The most obvious fact of North American stratigraphy is the way in which the main mobile belts run parallel to the margins of the present continent. This has often been contrasted with the situation on our side of the Atlantic where one can stand on the coast of south-west Ireland or Brittany and see the fold belts heading straight out into the ocean. In fact, in generalities, Europe and North America are remarkably alike. If one rotates one of them a little, everything falls into line. The Fenno-Scandian Shield and the Russian Platform match the Canadian Shield and the Stable Interior of the USA even to the 'syneclises' and 'anticlises' or 'basins' and 'domes'. The re-emergence of the basement in the Podolia Massif of the Ukraine is but a larger version of the Ozarks. The Urals become the Appalachians; the Volga, the Mississippi; and the Caspian Depression, the Mississippi Embayment. The really complicated parts of both continents are the western ranges of North America and the whole of western and central Europe. The big difference here is that while successive orogenies have operated along roughly parallel lines in North America, in Europe there have been two distinct trends; one parallel to the Atlantic, the other to the

Mediterranean. This, of course, relates directly to the plate histories of the two continents.

In Europe we have one major set of structures, from Precambrian to Recent, which relate first to continents splitting and colliding approximately along the Atlantic line and to the later sea-floor spreading that produced that ocean as we see it today. The second major set of structures start from the late Palaeozoic (though may also be much older) and relate to the collisions and less obvious splitting along the line of the Mediterranean/Tethys. In other words, the late Precambrian orogenies, the Taconian/Caledonian orogenies and the opening of the Atlantic are one thing, the Variscan and Alpine orogenies and the creation of the Mediterranean are another. The simplistic rule seems to be that almost everything in Europe (in modern terms) is either north-east/south-west or else it is east/west.

It has been well argued that development of the Atlantic began as far back as Precambrian times. The mobile belts of Proterozoic times extended the length of what is now the North Atlantic, from east Greenland and Norway in the north, via the British Isles, Newfoundland and the east coast of the United States to the west coast of Africa in the south. The plates were already defined and in motion. Later the system was stabilised and the present continents slowly built up and extended by plutonism.

At the end of Precambrian times, one cannot but be impressed by the similarities of deposits such as the Eocambrian of Greenland, the Swedish Jotnian, the Scottish Torridonian and the French Brioverian. These dominantly clastic sediments, with their glacial tillites, are almost everywhere followed by the quartzites, glauconitic sandstones and shallow-water limestones of the early Cambrian. Only locally do we seem to have an orogeny at this time, which is confused (as is so much else in stratigraphy) by nomenclatural anomalies. Thus the Czechs talk of an Assyntian orogeny, but there is no sign of such at Assynt in Scotland. A better name would seem to be Cadomian, after the Roman name for Caen in Normandy, where all the Precambrian sediments seem to have been tectonised before the beginning of the early Palaeozoic. The Avalonian orogeny of about the same age in Newfoundland, has been interpreted as involving the formation of a volcanic island arc and associated trench sediments on roughly north–south lines. It has been suggested that the serpentinites, glaucophane schists and associated rocks of Anglesey represent a subduction zone where oceanic crust was being consumed below the early European plate. The north-west Highland succession would then belong to the early American plate on the other side of an ocean of unknown width.

The same story would continue into the early Cambrian, where the contrasts between the north-west Scottish and the Anglo-Welsh successions are well known. If we include here that part of the Dalradian succession which is known to be Cambrian in age, then there would be an ocean of unknown width lost in the space now occupied by the Southern Uplands of Scotland.

In late Cambrian and early Ordovician times something was still clearly happening along the Atlantic line. A submarine trench has been postulated to accommodate the thick piles of Manx and Skiddaw slates in north-west England, perhaps superimposed on an earlier Cambrian island arc.

An early Ordovician trench is also thought to have extended from just north of the Southern Uplands of Scotland down across north-west Ireland. Along this trench, oceanic crust with ophiolites and 'turbidites' coming from the south-east was probably consumed down a Benioff zone. This in turn may have given rise to the high temperature metamorphism and calc-alkaline intrusions of the southern part of the Scottish Highlands and their prolongation in Ireland. It also probably gave rise to considerable north-westerly directed overthrusting in the Highland area, including the great Moine Thrust itself.

Ophiolitic associations of early Ordovician age occur throughout the Caledonide and Appalachian belts, and are thought to have originated as a result of sea-floor spreading between and within island arcs. In mid Ordovician times a volcanic island arc is thought to have extended through the English Lake District, Wales and Newfoundland. Perhaps at this point in time, there was a change from sea-floor spreading in the North Atlantic area, to sea-floor contraction. This may be reflected in the faunal changes and may have brought to an end the stable conditions represented by the pure quartzites that extend all the way from the Welsh borderland to Morocco. In fact, several of the early Palaeozoic facies and faunas seem to ignore the western Mediterranean and pass unhindered on into west Africa. This argues against many of the palaeomagnetic reconstructions that put them far apart. To me, north-west Africa is part of Europe and either the division is farther south or, if they must be sailing on different plates, then those plates were very close together long before the Alpine convulsions. It has been suggested that such sediments are characteristic of aseismic continental shelves as they move away from a mid-oceanic ridge.

This sort of evidence leads me to the conclusion that the continents, rather than sailing about the earth until they met in catastrophic collisions, separated and came together again repeatedly along the same general lines. In other words, there were many catastrophes and certain parts of each plate were particularly accident prone.

Meanwhile, back in the Ordovician, following on the stable episode, there were clearly major plate collisions, with the formation of oceanic trenches, Benioff zones and volcanic island arcs. Pre-Caradocian movements are widespread in eugeosynclinal sediments. The Taconian movements of eastern North America clearly represent the meeting of continents with the involvement of considerable quantities of oceanic material. The same movements are probably much more in evidence in western Europe than we usually seem willing to admit. Orogenies and accompanying metamorphism of about this age (that is, about 478 million years BP) have been recognised, for example, in the Massif Central, the Vosges and the Black Forest. Following these movements we have a marked regression followed by the widespread transgressions, both in Europe and North America, of the late Llandovery and Wenlock. So by mid Silurian times the plates that are now in the northern hemisphere were again stable and low in relief. It would seem that the plates which had come together in the Taconian orogeny, with the subduction of a proto-Atlantic plate and the westerly overriding of the sedimentary pile along the line of the Appalachians had now more or less stopped. The Wenlock/Niagaran limestones spread gently and widely on the shallow shelves. The lack of provincialism in Silurian faunas probably reflects this stabilisation of oceanic plates that had been temporarily welded together. Volcanicity had been again reduced to a minimum, though what there was showed itself very significantly along the line of the future Variscan front. This may, in fact, be the first whisper of the Tethys.

The Caledonian orogeny in reaching its climax at the end of Silurian times, expressed itself along the same Taconian lines and along the lines that were, much later, to be the Atlantic Ocean. Though evidence of the Caledonian orogeny has been claimed from areas as remote as the Canadian Rockies and the Bohemian Massif, it is fairly obvious where its main effects were felt. It is, perhaps, significant that the long search for the ideal section to define the Silurian/Devonian boundary went from Czechoslovakia to the Ukraine, to Morocco and to the western regions of the USA and Canada. Here not only do the sediments preserve a continuous and more or less uniform record over the disputed boundary, but the faunas (most notably the mono-graptids) went on happily evolving, oblivious of the great things that were going on elsewhere.

One must presume that, as a result of the Taconian and Caledonian orogenies, the American and European plates were now close together. In fact the amount of oceanic crust involved in the rocks of the Caledonides suggests that a short-lived proto-Atlantic had come to an end and the two continental shelves had met. This may explain the evident uniformity of Silurian facies and Silurian faunas. There followed the post-orogenic sediments of the Old Red Sandstone, extending from Canada to Asia. Perhaps it was the orogeny with its crustal shortening in the continents that produced the widespread marine regressions as suggested earlier in this chapter. Coupled with possible pulling apart, this may have produced the marked provincialisation recognised in Devonian marine faunas.

By mid and late Devonian times, the new mountains of Europe had been worn down and shallow shelf seas spread on to the continental margins. In effect it was a return to the conditions of mid Silurian times, with a corresponding decrease in provincialism, but it lasted longer and led to the fantastic flowering of the Frasnian reefs. No doubt there were also climatic factors involved and one automatically looks to the reconstruction maps to see if all the reef-growing areas, from Canada to Australia, were placed in the right latitudes. Not surprisingly, it seems that they were.

Now we come to one of the great anomalies of the stratigraphical record, with the widespread extinctions of the Frasnian/Famennian junction. There is no evident explanation to be found in drifting plates or colliding continents. It seems that here, at least, we must appeal to an extra-terrestrial cause. It must be remembered, however, that there were important earth movements in North America in mid Devonian times. These are the Acadian movements of eastern Canada, where folded Early Devonian rocks are overlain with marked unconformity by Late Devonian sediments. That these movements were still essentially Atlantic in their orientation, is shown by the great accumulations of syn-orogenic and post-orogenic sediments down through the Appalachian belt of the United States. Of these, the most famous are the vast Late Devonian accumulations of the Catskill Mountains.

The Early Carboniferous shelf seas which followed the Late Devonian regression were really, in their generalities, very like the Givetian and Frasnian seas of the Devonian. The big difference was that caused by the earlier widespread extinctions. The reefs and all the other marine biota, had to be constructed from different organic building blocks. The stromatoporoids, for example, had largely gone and the new corals had not yet learned to build reefs (in fact the Rugosa were never to remember the habit and it was only to be relearned by the Scleractinia of the Mesozoic). The Waulsortian and other reefs of the Carboniferous were largely founded on bryozoans and algae. Others (including the 'type' bioherms in Indiana) were nothing more than heaps of crinoid debris.

Following on from the suggestion of plate separation in the Devonian, the Early Carboniferous was characterised by tension and rifting, as in the Midland Valley of Scotland, with its extensive lava flows. In the Appalachian region there were repeated uplifts giving rise to coarse detritus in the trough to the west, whilst a vast shallow carbonate-depositing sea extended over the stable interior of North America as far as the Pacific ranges.

We know, however, that preliminary rumbles of the Variscan orogeny were already beginning to be heard at the end of Dinantian times. In North America uplift and erosion at this time were enough to divide the Mississippian and Pennsylvanian periods, at least in their nomenclature. What this meant in terms of plates is not clear, but one can see in Europe that something fundamental had happened and we are now dealing, not with an Atlantic line, but with a Mediterranean one. It is possible that this too was the resurrection of a much older trend, possibly as old as the Hudsonian–Ketilidian–Laxfordian–Svecofennian orogeny of Archaean times.

From palaeomagnetic evidence, it is alleged that the African and European continents moved vast distances during the Devonian and only began to approach each other during Carboniferous times. I am a little unhappy about this, because there seem to be such close resemblances between the two plates, that I would not wish them ever to be far apart. Alternatively, did the Tethys/Mediterranean, like the Atlantic, open and shut several times? Was the Alpine orogeny no more than an encore for the Variscan performance? Whatever the process, it could well be that the crustal shortening in the continents produced by the early phases of the Variscan orogeny, with its resultant lowering of sea-levels, produced the widespread regression of Late Carboniferous times, when coal measure swamps spread from Texas to the Donetz.

The main onslaught of the Variscan orogeny, as seen in southern Europe, occurred in Mid–Late Carboniferous times, before the deposition of Stephanian sediments in the inter-montane basins. Thus in Cantabria, in northern Spain, there is a clearly displayed angular unconformity at this level, followed by thick conglomerates. In so far as it can be seen through the confusion of the Alpine orogenesis, this situation is wide-spread in the Mediterranean lands. Farther north in Europe, the later Carboniferous is marked by a progressive drying out and loss of marine influence. It is noteworthy that movements in Europe seem to coincide with a break within the Pennsylvanian System in North America, even though there are no traces or orogenic episodes in that continent.

The main orogeny in the north of Europe, coming at the end of the Carboniferous, is defined by the so-called 'Variscan front'. This line can be traced, rather precisely, from New England, through south-west Ireland, via south Pembrokeshire and the Gower Peninsula, under the University College of Swansea, then south of the Kent coalfield to the Boulonnais and on as the *Grande Faille du Midi* far into the European continent. It is the earlier line of Silurian volcanicity and also roughly the line of change from marine Devonian facies to the south and continental Old Red Sandstone facies to the north. It is Stille's boundary between 'Palaeo-Europe' and 'Meso-Europe', with gently-folded late Palaeozoics to the north and strongly-folded late Palaeozoics to the south. It may also be significant that the east–west structures of southern Britain become north–south to form the Pennine backbone of northern England. In other words, they swing from a Mediterranean to an Atlantic orientation. It must also be remembered that if one swings Spain (as one must) to close the Bay of Biscay, then the main Variscan trend of the Iberian Meseta becomes an Atlantic rather than a Mediterranean direction. The 'Rodilla Asturica' or Asturian kneebend then represents a swing from the latter to the former. Similarly the Variscan folding of the Anti-Atlas in southern Morocco swings from an 'Atlantic' direction in the west to the 'Mediterranean' direction farther east.

It has been suggested that starting in Mid Devonian times and continuing on through the Carboniferous, a mid European ocean of uncertain width extended roughly along the line of the English Channel and then on eastwards into the European continent. Certainly there are some little puffs of pillow-lavas and associated rocks of this age which may represent subduction zones going down under southern Britain, the Rheinisches Schiefergebirge and the Harz on one side and under Armorica and the Vosges on the other. But they are nothing compared with the Caledonian and Alpine ophiolite suites. What is more, there is very little that can be called oceanic crust or trench-filling sediment (flysch, olistostromes, etc.) within the Variscan massifs of Europe. Perhaps it is all hidden under Alpine fold-belts. Personally, I prefer the notion that this structure, whatever it was, did not open very far at this time and soon closed again. What seems to me more significant is that it both foreshadowed and paralleled the Variscan, Tethyan and Mediterranean lines that were to come.

The general drying out of northern Europe continued through the Permian, with significant interruptions such as the Zechstein Sea (when Carboniferous Limestone type conditions tried to reassert themselves). Whether this was a continuation of the regression story or in part the creation of a rain-shadow area behind the new mountains, it is difficult to deduce.

The great compressional movements of colliding continents subsided and were replaced by the vertical movements of tension and isostatic readjustment. Also at this time came the great compressional movements of the Urals. Here also there was a much older history of troughs and orogenies, most obviously the late Precambrian and Palaeozoic trough of Timan, which makes a narrow angle with the northern Urals, much in the same way that the 'Palaeo-Rockies' make one with the later Rockies. In the Urals the late Palaeozoic geosyncline persisted through the greater part of the Permian and the Asian record is, on the whole, very different from that of Europe.

This great structure, which forms the frontier between Europe and Asia, shows two continents coming together and forming a new super-continent. Perhaps, as with the Atlantic and the Mediterranean, this is where we may expect the next rifting, though I doubt if it will happen soon enough to substantiate my idea.

The pattern in north-west Europe was now very much one of tension and rifting. The Vale of Eden between the Pennines and the Lake District in northern England provides a small but good example of this happening in Permian times. The extrusion of great quantities of quartz porphyry in the Oslo Graben and in the South Tyrol and of other volcanics in the Black Forest may be further examples. In the Massif Central of France there were a whole series of little grabens and the remarkable *sillon houiller*, a rift structure of this age, only about 2 km wide.

Meanwhile, in North America, the orogeny that produced the Appalachian structures probably did not reach its climax until the end of the Permian. The great compressive forces were again operative along the Atlantic line and more than 320 km of crustal shortening has been estimated. Granitic batholiths welded the geosynclincal rocks on to the continent.

It must not be forgotten, however, that in the south of the United States there are roughly east–west structures such as the Ouachita Mountains which may relate to a Palaeo-Tethys and be the equivalent of the Variscan fold belts of Europe. It must also not be forgotten that we in Britain, at least, tend to get our latitudes wrong when looking across the Atlantic. We should not try to match New England with Old England. Before separation, what was to be the USA was close against southern Europe and northern Africa, and it is only the Gulf Stream that makes our Labradorian latitudes bearable.

Following on from the evidence of tension in Permian Europe, in the Trias there seem to have been rifting structures everywhere, such as for example, the horst and graben structures of the Midland coalfields of England. Recent evidence from boreholes clearly demonstrates the control of Triassic basin sedimentation by marginal faults. The great graben of the Vale of Severn and the Cheshire Basin may be thought of as an early (prenationalist) attempt at separating Wales from England. But writing the first edition in the midst of the Common Market discussions of 1972, it seemed appropriate to remark that Britain decided some 200 million years ago to remain with Europe rather than depart with North America.

The close resemblance of the Newark Group along the eastern seaboard of the USA to the Triassic deposits of western Europe has already received comment. Equally striking is the similarity of the structures in which these sediments are preserved. These are perhaps the best American examples of Marshall Kay's 'taphrogeosynclines', that is to say, geosynclines with faulted margins. It is relevant to the theme of this chapter to point out that they have also been called 'epieugeosynclines', i.e. fault-bounded geosynclines developed on top of old eugeosynclines. The significant point here is again the coincidence and parallelism of structures along the New England coast. Though many of us would not wish to dignify such phenomena with the name 'geosyncline', they are certainly very thick accumulations in fault structures roughly parallel with the Atlantic coast. There seems little doubt that they mark the tensional effects of the splitting apart of the two Atlantic-facing plates. Probably there had been an earlier Atlantic which came to an end with the Caledonian orogeny, but this was the beginning of the ocean that we have today.

It is appropriate at this point to digress for a moment on the subject of the faunistic changes that took place at the end of the Palaeozoic. This was one of Norman Newell's great catastrophes in the history of life (though it is important to note that the floristic changes were to come later). In one of his later articles, Professor Newell has commented: 'Of these great changes (in the physical environment) no single set of factors has influenced the later history of life more than the continuous modifications of topographic relief and distribution of continents and ocean basins and the attendant climatic changes.' I subscribe whole-heartedly to these views and would not in fact restrict them to the 'later history'. Major geographical changes appear to have caused widespread extinctions by the destruction of habitats. This in turn has led to rapid evolution to fill the vacant or new ecological niches. As H. L. Hawkins once wrote: 'Death makes room for life.' I am coming more and more to the view that the evolution of life, like the evolution of continents and of the stratigraphical column in general, has been a very episodic affair, with short 'happenings' interrupting long periods of nothing much in particular.

Much is hidden in the mists of the early Triassic, which is probably the least-known episode in the long history of Phanerozoic time. Nevertheless, one may reasonably presume that the major changes in the marine faunas at this time followed from the regressions caused by the Variscan orogeny. These are widely advertised in the evident continentality of so much of the Late Permian and Early Triassic record in Europe and North America. The welding together of plates in the orogeny would have tended to preserve the equanimity of the land floras and the land vertebrates. The plants at least only showed significant changes with the rifting of the plates in later Triassic times.

The one feature that clearly differentiates between the Newark Group of eastern North America and the classic Trias of north-west Europe, is the presence of basic igneous intrusions and extrusions in the former. It is therefore significant that this feature is also seen in the Trias of Spain and Morocco, which, according to the most generally-accepted plate reconstructions, would have been close to the eastern seaboard of the States. These were therefore probably closest to the main line of plate rifting. The other graben structures in western Europe, already discussed, were presumably 'tentative' splits which never came to anything.

So the Atlantic was defined in the Triassic and, in fact, there is increasing evidence to suggest that much of the shape of western Europe was blocked out in Mesozoic times. Many of the arcuate lengths of contemporary coastlines now seem to be the edges of Triassic basins. It has been said that the Irish Sea and even the Bristol Channel (just outside my window as I write) are older than the Atlantic. This is true in the sense that Mesozoic sedimentary troughs and basins developed and were complete in these areas before the main opening of the North Atlantic. Some of the marginal faults—notably that along the Sutherland coast discussed earlier—were clearly operative during Mesozoic times (in this case late Jurassic) and may well have been active seismic lines. Others, such as the fault that let down the great thickness of Mesozoic in Cardigan Bay (to the west of Wales) must have been much later, though they served to define basins that were presumably already there. The marginal Jurassic strips of east Greenland are similarly fault-bounded. It is surely significant (both scientifically and economically) that the great 'hard rock' massif of Scotland is now known to be almost entirely surrounded by much later 'soft rock' basins.

Apart from these tensional effects, most of Mesozoic times over most of Europe may be thought of as what I have called the 'long quiet episode'. Through the greater part of Triassic, Jurassic and Cretaceous times, what is now northern Europe experienced a remarkably uneventful history that made possible the gentle (if spasmodic) accumulation of sediment and the slow evolution of organisms which led to the growth of the science of stratigraphical palaeontology in these strata. For large areas there is not so much as a pebble bed to make one stumble in the climb up the column. In this passivity, the infancy of stratigraphy may have brainwashed us into thinking that this is the true nature of the stratigraphical record. The same is true in eastern and southern North America, though much of that record is out on the continental shelf. Elsewhere (as along the western seaboard of North America) plates were converging and being subducted, but over a great part of the earth's surface, and my part of it in particular, the plates were gently moving apart as the Atlantic formed from its median ridge. For most of this time, for example, much of northern Europe was in the condition of a shallow, flooded shelf. The incredible persistence of the 'Blue Lias' facies of Early Jurassic times is symptomatic of this. It is no coincidence that the simple old German sub-division of the Jurassic into 'Black Jura', 'Brown Jura' and 'White Jura' is still so effective around Europe.

Away in the south of Europe, as in western North America, there are flysch deposits, the ophiolite suite and much else besides, indicative of island arcs, deep trenches and ocean floors.

The quiescence of the Jurassic culminated in the widespread Tithonian limestones of southern Europe. Another feature of the Jurassic in southern Europe is the *Rosso ammonitico* facies of nodular red limestones characterised by a dominance of pelagic organisms. On its own it has been explained in various special ways, but considered in the general setting of Mediterranean stratigraphy, some more general explanation seems to be needed. Deposits very similar to the Jurassic *Rosso ammonitico* occur at several levels on both sides of the Mediterranean. In the Palaeozoic they are usually called *griottes* from their supposed resemblance to cherries, and can be found (in northern Spain and southern France, for example) in the Cambrian, the Devonian and the Upper Carboniferous, besides their very wide distribution in the Jurassic. The fact that they also occur in Algeria and Tunisia, for example, implies similar depositional environments and probably no great separation. It is difficult to avoid a climatic explanation for this facies, just as the differences between the Tithonian and the contemporaneous Volgian facies farther north were also probably climatic rather than physiographic.

But the most important conclusion one must reach about Jurassic times is that much of the form of the present continents had then been blocked out as it is today. This point was well expressed by the great Arkell shortly before he died. He wrote: 'All the occurrences of Jurassic formations . . . amount to little more than relics or marginal lappings of the sea around the edges of the continents; the sole exception being the Tethys.'

Passing on now to the Cretaceous, there were complications at the beginning of that period, perhaps related to the poorly dated orogenies of other parts of the world. Nevertheless, these were not very significant and some facies—notably the 'Wealden' of western Europe—are again remarkably widespread. The 'Urgonian' was, in its way, very similar to the Tithonian. In the French Jura, for example, they are often remarkably alike in facies and fauna. But something very serious happened in Mid Cretaceous times after the deposition of the 'Urgonian'. If we pursue the doctrine of the orogeny/regression couple, then we must also expect the corollary that the wearing down of mountain ranges and the spread of continents by marginal sedimentation must lead to widespread transgressions. Certainly at about the beginning of Late Cretaceous times we have one of the best documented transgressions of all time. The 'Cenomanian transgression' turns up nearly everywhere—all over Europe and northern Africa and round the Mississippi embayment of North America, and in many other places besides. I should perhaps say the 'so-called Cenomanian transgression' for it often differs in age by a stage or so, but nevertheless, it was the most startling event of Cretaceous times. It cannot be better exemplified than by the unconformable relationship of late Cretaceous marine sediments to the metamorphosed Precambrian rocks of the Bohemian Massif, as illustrated in Figure 3.1.

If my theorising is correct, then the 'Cenomanian transgression' was in part due to wearing down of the Variscan and perhaps later ranges (as in the western United States) together with marginal sedimentation. It may be compared with the Early Cambrian transgression after the Cadomian orogeny, the mid Silurian transgression after the Taconian orogeny and the Early Carboniferous transgression after the Caledonian and Acadian orogenies. But this is an obvious over-simplification, and I would not wish to go back on my words earlier in this chapter about those wise men who claim to see a pattern in history. Unless and until we find a deterministic control, presumably deep inside the earth, I can see only a series of accidents; most of these accidents seem to be the collisions of continents on the earth's surface. It has recently been argued very convincingly that the Cenomanian transgression can be blamed on a pulse of rapid ocean-floor spreading with the expansion of the mid-ocean ridges and consequent displacement of water.

The biggest complications of mid Cretaceous times were the circumstances that produced the first great spasm of the Alpine orogeny. The importance of the Mid Cretaceous movements is becoming increasingly recognised. They certainly played a major part in the history of the eastern Alps, the Carpathians and the Balkan Mountains, with widespread northerly thrusting of nappes. It was also at this time that there occurred the 'inversion' of Voigt's *Randtröge* or marginal troughs in northern, extra-Alpine Europe. As a result, older ridges (such as the London–Brabant Massif) became buried under the sediments of new, secondary troughs.

One can only deduce that the Eurasian and African plates began driving together at this time, with the latter dragged down beneath the rising Alpine mountains. This would seem to contradict, however, the notion of major transgressions being the much-delayed after-effects of an orogeny.

The remarkable stability of northern Europe and eastern North America after this mid Cretaceous spasm is surely clear evidence of the episodic nature of some of the plate movements. The persistence of the white chalk facies outside the new mountain belts must clearly indicate a long period when the Eurasian and American plates were not being affected by any major continental collisions. It still remains difficult to explain the Gingin Chalk of faraway Australia, though one might guess that this is part of the same phenomenon that produced the almost world-wide 'Cenomanian transgression'. But we must not forget the great flysch troughs which were developed in Alpine Europe through much of Cretaceous time and which in many cases continued on into the Tertiary.

By this time, the southern part of the north Atlantic was presumably wide open, and I strongly suspect a crack going up as far as east Greenland. I have earlier suggested that the presence of Tethyan elements (of Late Jurassic and Early Cretaceous age) in Greenland might be explained in terms of an early Gulf Stream sweeping its way into an incipient North Atlantic. So might also the presence of other Mediterranean genera and species in the more westerly outcrops of Britain.

The main opening of the North Atlantic, however, was evidently a Tertiary affair. The volcanic history of that event does not need restatement here, but the complexity of the stratal history makes this part of the column both the most confusing and the most controversial of all. The widespread regression at the end of Cretaceous times may be related to three major plate phenomena:

   (i) the Laramide orogeny along the western edge of the American plate,

 (ii) the opening of the northern part of the North Atlantic, and

(iii) the further grinding together of the African and Eurasian plates to produce the early Tertiary Pyreneean folding of Cantabria, the Pyrenees and Provence in south-west Europe.

New east–west fold mountains immediately began to be destroyed again, producing mountains of conglomerate such as the fantastic shapes of Montserrat, near Barcelona (Figure 8.3) which is remarkable even in a country of conglomerates like Spain. The Atlas folding in north-west Africa conforms remarkably with that of the Pyrenees, which provide probably the best evidence we have in Europe of the coincidence of Variscan and Alpine fold belts.

The Variscan massifs of Europe, such as the Spanish Meseta, the Massif Central of France, the Eifel of Germany, the Bohemian Massif of Czechoslovakia and the Rhodope Massif of Bulgaria are characterised by roughly north–south tensional grabens filled with Tertiary non-marine sediments. They also display all the features of a volcanicity that lasted late enough to terrify Palaeolithic man and perhaps to provide him with his fire. Some of the craters look so fresh that one almost expects the rocks still to be warm. But this volcanicity also testifies to the great tensional stresses that were still affecting Europe, producing vast fissure eruptions and long straight lines of volcanic cones, such as the Chaine de Puys near Clermont Ferrand (Figure 8.4) and the puy-like volcanoes in the 'Czech Auvergne' on the west side of the Bohemian Massif (Figure 8.5).

Figure 8.3 *Montserrat, near Barcelona, Spain; a strangely shaped mountain of Miocene conglomerate (DVA)*

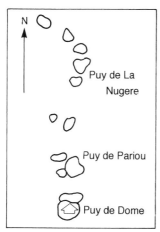

Figure 8.4 *La Chaîne des Puys, a line of late Caenozoic volcanoes along a fracture line in the Auvergne, seen from the top of the Puy de Dôme, near Clermont Ferrand, France (DVA)*

The main spasm of the Alpine orogeny came in Oligocene and early Miocene times, with the African plate grinding against Europe and the alpine chains spreading out to the north and south like an opening fan. So much is now known of the Alpine fold belts, the times and forms of their movements, and so much is now being deduced about the relationship of all this to the theories of plate tectonics, that I marvel at my audacity in saying anything at all at this stage. Clearly it is more than just a picture of two continent-carrying plates coming together. As with every theory that seems at first to have an obvious answer to every problem, the beautifully simple picture is

Figure 8.5 *Puy-like volcano near Karlovy Vary (formerly Karlsbad) on the Bohemian Massif of Czechoslovakia (DVA)*

becoming smudged; the two obvious large plates of Africa and Europe are being confused by the 'microplates' of the eastern Mediterranean. The most apparent anomaly is that the Mediterranean, although it has in Cyprus something comparable to a mid-oceanic ridge, does not have the magnetic 'striping', with reversals, of other oceans. The eastern Mediterranean is still evidently an active tectonic area of colliding plates, with Crete and adjacent ridges as part of a small island arc parallel to the Hellenic trench.

The western Mediterranean, however, is quite different, having alpine folds to the south as well as to the north. Here, perhaps, all the ocean-floor material has been carried up into the mountains. Flysch-bearing troughs and ophiolites are major constituents of the allochthonous elements in the Alpine chains, including those in the Apennines discussed earlier.

A similar belt of Mesozoic ophiolites has been traced from the Taurus Mountains of southern Turkey, just into northern Syria and Iraq and then on through Iran to Oman. These appear to relate to an earlier phase of ocean spreading during Late Triassic times, and were carried to their present positions in nappes formed during the Cretaceous, as in eastern Europe.

One feature of modern plate systems that does seem to show fairly well in the western Mediterranean is the transform fault. Most notably the Sestri–Voltaggio line separates the reverse-facing French Alps and Italian Apennines. In the former the tectonic movement and the migration of troughs is to the south and south-west. In the Apennines, it is towards the north-east. Comparable units in the two ranges seem to have come from an oceanic trough that was torn apart and thrust in opposite directions. It has been suggested that, though the transform fault itself has disappeared, the Sestri–Voltaggio line indicates its former position. It separated Benioff zones that were dipping and consuming oceanic crust in opposite directions, as in the south-west Pacific trench system at the present day. Other wrench faults around the Mediterranean may be interpreted similarly, most notably between the eastern Alps and the Dinarides, where similarly opposing structures are found in close proximity.

The last push of the Alpine mountains over their own Miocene molasse-detritus is now known in all their constituent ranges. Since then the story in Europe has been essentially one of settling down after the storm. At first the new mountains must have been worn down with great rapidity to produce the vast quantities of Neogene conglomerates. Then there were the molasse-type sands, which in their drab, khaki featurelessness are recognisable in the smallest exposures from Spain to Bulgaria and beyond.

The later shelly sands of the northern Neogene (such as the English 'crags' and the French *faluns*) and the echinoid-bearing limestones of the Mediterranean may represent a return to plate stability. We must remember, however, that just as all the Alpine chains of Europe are now known to have been still pushing forward over the molasse in Late Miocene times, so in places such as the Apennines, movements were still going on as late as Quaternary times. In the Californian geology that is so much unlike everyone else's, there are right-angled unconformities within the Pleistocene, so the plates there were certainly still moving. And we know, of course, that the Atlantic is still tearing along the dotted line down its centre, albeit very slowly in human terms. Even as I wrote the final corrections to the first edition of this text in January 1973, a splendid, though destructive new tear had opened across the island of Heimaey, south of Iceland, and new oceanic crust was being formed. The tensions down the east side of Africa are well known and my colleague Peter Styles has shown that in the Red Sea the geophysical evidence also indicates episodicity.

It is also worth noting here that the only Mid Tertiary movements in northern Europe that can be called an orogeny are in Spitsbergen, where we find the last expression of an Atlantic compressive line clearly preserved down the west coast, presumably reflecting the continued grinding together of the northernmost tip of the European plate with that of Greenland.

The events of the Pleistocene glaciations must never be overlooked in any consideration of the stratigraphical record. Those glaciations were the most obviously catastrophic events in our history and produced in their tills and periglacial deposits some of the most persistent facies of all. In our near-sighted way of looking at the stratigraphical column, we tend to forget that these recent events, if considered on the normal geological time-scale, were virtually instantaneous and certainly catastrophic. The whole of the Pleistocene ice age would fit within an ammonite zone or two.

The final conclusion I come to therefore is that, though the theories of plate tectonics now provide us with a *modus operandi*, they still seem to me to be a periodic phenomenon. Nothing is world-wide, but everything is episodic. In other words, the history of any one part of the earth, like the life of a soldier, consists of long periods of boredom and short periods of terror.

REFERENCES

Ager, D. V. (1963). *Principles of Paleoecology*. McGraw-Hill, New York. For brief discussions of various faunal and flora distributions in the fossil record, including the suggestion of a hypertropical zone (p. 156).

Ager, D. V. (1975). 'The Geological Evolution of Europe', *Proc. Geol. Ass., Lond.*, Vol. 86, pp. 127–154. A more expanded (albeit still ridiculously condensed) account of the history of Europe seen in modern terms.

Ager, D. V. (1980). *The Geology of Europe*. McGraw-Hill, Maidenhead, 535 pp.
A perhaps reckless attempt to summarise the geology of a wonderful continent.

Audley-Charles, M. G. (1970). 'Stratigraphical Correlation of the Triassic Rocks of the British Isles', *Quart. J. Geol. Soc.*, Vol. 126, pp. 19–47.
A beautiful demonstration of the control of Triassic sedimentation by tensional faulting.

Berry, W. B. N. (1972). 'Early Ordovician Bathyurid Province Lithofacies, Biofacies and Correlations—Their Relationship to a Proto-Atlantic Ocean', *Lethaia*, Vol. 5, pp. 69–88.
Early Ordovician plate movements deduced from faunal changes and tectonics.

Dewey, J. F. (1971). 'A Model for the Lower Palaeozoic Evolution of the Southern Margin of the Early Caledonides of Scotland and Ireland', *Scot. J. Geol.*, Vol. 7, pp. 219–240.
An interesting paper reinterpreting classic country in the light of the ideas of plate tectonics.

Dewey, J. F. and Bird, J. M. (1970a). 'Mountain Belts and the New Global Tectonics', *J. Geophys. Res.*, Vol. 75, pp. 2625–2647.

Dewey, J. F. and Bird, J. M. (1970b). 'Plate Tectonics and Geosynclines', *Tectonophysics*, Vol. 10, pp. 625–638.
Two already classic papers on the new geology with its repercussions (not yet by any means fully worked out) on the understanding of the stratigraphical record.

Evans, G. (1970). 'Coastal and Near-Shore Sedimentation: A Comparison of Clastic and Carbonate Deposition', *Proc. Geol. Ass.*, Vol. 81, pp. 493–508.
An excellent account of Recent coastal sedimentation in tropical and temperate climates.

Gass, I. G., Butcher, N. E., Clark, P., Drury, S. A., Francis, P. W., Jackson, D. E., McCurry, P., Skipsey, E., Smith, P. J., Stevenson, J., Thorpe, R. S., Turner, C., Wilson, R. C. L. & Wright, J. B. (1972). *Historical Geology*. Open University Press, Bletchley, 64 pp.
The most refreshing breath of fresh air we have had in British stratigraphy for a very long time.

Girdler, R. W. and Styles, P. (1974). 'Two Stage Red Sea Floor Spreading', *Nature, Lond.*, Vol. 247, pp. 7–11.
A clear demonstration of episodicity in ocean-floor spreading.

Grabau, A. W. (1937). 'Fundamental Concepts in Geology and Their Bearing on Chinese Stratigraphy', *Bull. Soc. Geol. China*, Vol. 16, pp. 127–176.
Useful for a brief statement of his 'pulsation theory' of earth history.

Grasty, R. L. (1967). 'Orogeny, A Cause of World-Wide Regression of the Seas', *Nature, Lond.*, Vol. 216, pp. 779–780.
A brief paper with a very good idea.

Hallam, A. (1963). 'Eustatic Control of Major Cyclic Changes in Jurassic Sedimentation', *Geol. Mag.*, Vol. 100, pp. 444–450.
It's all done by sea-levels.

Hays, J. D. and Pitman, W. C. (1973). 'Lithospheric Plate Motion, Sea Level Changes and Climatic and Ecological Consequences', *Nature, Lond.*, Vol. 246, pp. 18–22.
A quantitative demonstration that the Mid Cretaceous transgression and subsequent regression may have been caused by a pulse of rapid ocean-floor spreading which then slowed down again.

Hollin, J. T. (1962). 'On the Glacial History of Antarctica', *J. Glaciology*, Vol. 32, pp. 173–195.
In which the fluctuations of the Antarctic ice-cover are shown to relate to those of the northern ice sheets, thence implying a world-wide control. Also the suggestion of seeking the cause in the stratigraphy of Mars.

Hollin, J. T. (1969). 'Ice Surges and the Geological Record', *Can. J. Sci.*, Vol. 6, pp. 903–910.
Sudden surges of polar ice producing very widespread effects.

Hughes, C. J. (1970). 'The Late Precambrian Avalonian Orogeny in Avalon, South-East Newfoundland', *Am. J. Sci.*, Vol. 269, pp. 183–190.
An interesting account of this orogeny and the elongated island arc and trough formed here at this time.

Ma, T. Y. H. (1954). 'Climate and the Relative Positions of the Continents During the Lower Carboniferous Period', *Acta Geol. Taiwanica*, No. 6, pp. 1–86.
One of a whole series of papers relating to the distribution of Palaeozoic coral growth forms to an equator that requires drifting and polar wandering.

Mitchell, A. H. and Reading, H. G. (1971). 'Evolution of Island Arcs', *J. Geol.*, Vol. 79, pp. 253–284.
Island arc theory applied to the stratigraphical record, especially that of the early Palaeozoic on either side of the North Atlantic.

Meyerhoff, A. A. (1970). 'Continental Drift II. High-Latitude Evaporite Deposits and Glacial History of Arctic and North Atlantic Oceans', *J. Geol.*, Vol. 78, pp. 406–444.
As an example of the use of sediment distributions as ammunition against drifting hypotheses.

Newell, N. D. (1971). 'Faunal Extinction', in *Encyclopedia of Science and Technology*, 3rd edition. McGraw-Hill, New York.
A semi-popular summary of Professor Newell's views.

Ronov, A. B. (1959). 'On the Post-Precambrian Geochemical History of the Atmosphere and Hydrosphere', *Geochem.*, No. 5, pp. 493–506. Correlation of volcanicity and carbonate deposition.

Scholle, P. A. (1970). 'The Sestri-Voltaggio Line: A Transform Fault Induced Tectonic Boundary Between the Alps and the Apennines', *Am. J. Sci.*, Vol. 269, pp. 343–359.
An interesting example of how plate tectonic theory is being used to explain the Alpine mountain systems of Europe.

Sloss, L. L. (1963). 'Sequence in the Cratonic Interior of North America', *Bull. Geol. Soc., Amer.*, Vol. 74, pp. 93–114.
The recognition of six major regressions on the North American craton, with a seventh going on at the present day.

Sloss, L. L. and Speed, R. C. (1974). 'Relationships of Cratonic and Continental-Margin Tectonic Episodes', in W. R. Dickinson (ed.), *Tectonics & Sedimentation, Spec. Publ. Soc. Econ. Paleont. Mineral.*, No. 22, pp. 98–119.
Application of the same principle to cratons generally, with submergent and oscillatory episodes in between, all related to plate tectonics.

Spjeldnaes, N. (1961). 'Ordovician Climatic Zones', *Norsk. Geol. Tidsskr.*, Vol. 41, pp. 45–77.
As an example of the use of sediment distributions (in this case Lower Palaeozoic carbonates) and shell characters as evidence in favour of drifting hypotheses.

Stehli, F. G. (1957). 'Possible Permian Climatic Zonation and Its Implications', *Amer. J. Sci.*, Vol. 255, pp. 607–618.
An important paper which, though much criticised, has led to a great deal more constructive thought.

Sutton, J. (1968). 'Development of the Continental Framework of the Atlantic', *Proc. Geol. Ass.*, Vol. 79, pp. 275–303.
A brilliant presidential address written just before the full onslaught of plate tectonics burst upon us.

Walton, E. K. (1970). 'Geosynclinal Theory and Lower Palaeozoic Rocks in Scotland', Tomkeieff Memorial Lecture, University of Newcastle-upon-Tyne.
A delightfully simple exposition of the latest ideas.

Wilson, J. T. (1966). 'Did the Atlantic Close and Then Reopen?' *Nature*, Vol. 211, pp. 678–681.
The suggestion on the basis of facies and faunas that a Proto-Atlantic closed to form the Appalachians and then reopened.

Index